CAMBRIDGE LIBRARY COLLECTION

Books of enduring scholarly value

Earth Sciences

In the nineteenth century, geology emerged as a distinct academic discipline. It pointed the way towards the theory of evolution, as scientists including Gideon Mantell, Adam Sedgwick, Charles Lyell and Roderick Murchison began to use the evidence of minerals, rock formations and fossils to demonstrate that the earth was older by millions of years than the conventional, Bible-based wisdom had supposed. They argued convincingly that the climate, flora and fauna of the distant past could be deduced from geological evidence. Volcanic activity, the formation of mountains, and the action of glaciers and rivers, tides and ocean currents also became better understood. This series includes landmark publications by pioneers of the modern earth sciences, who advanced the scientific understanding of our planet and the processes by which it is constantly re-shaped.

Illustrations of the Viscous Theory of Glacier Motion

This book brings together works published between 1846 and 1859 by the Scot James D. Forbes (1809–68) and Irishman John Tyndall (1820–93), both of whom were experienced alpinists as well as glaciologists. However, their views on the motion of glaciers were disparate, and a scientific quarrel over primacy and credit for discoveries continued even after their respective deaths. These papers include Forbes' articles on experiments on the flow of plastic bodies and analogies between lava and glacier flows, and on the plasticity of glacier ice, as well as Tyndall's observations on the physical phenomena of various Alpine glaciers, including the famous 'Mer de Glace', and a piece on the structure and motion of glaciers, co-written with Thomas Huxley. Several works by and about all three scientists (including works on Alpine travel) have also been reissued in this series.

Cambridge University Press has long been a pioneer in the reissuing of out-of-print titles from its own backlist, producing digital reprints of books that are still sought after by scholars and students but could not be reprinted economically using traditional technology. The Cambridge Library Collection extends this activity to a wider range of books which are still of importance to researchers and professionals, either for the source material they contain, or as landmarks in the history of their academic discipline.

Drawing from the world-renowned collections in the Cambridge University Library and other partner libraries, and guided by the advice of experts in each subject area, Cambridge University Press is using state-of-the-art scanning machines in its own Printing House to capture the content of each book selected for inclusion. The files are processed to give a consistently clear, crisp image, and the books finished to the high quality standard for which the Press is recognised around the world. The latest print-on-demand technology ensures that the books will remain available indefinitely, and that orders for single or multiple copies can quickly be supplied.

The Cambridge Library Collection brings back to life books of enduring scholarly value (including out-of-copyright works originally issued by other publishers) across a wide range of disciplines in the humanities and social sciences and in science and technology.

Illustrations of the Viscous Theory of Glacier Motion

JAMES D. FORBES

CAMBRIDGE
UNIVERSITY PRESS

University Printing House, Cambridge, CB2 8BS, United Kingdom

Cambridge University Press is part of the University of Cambridge.
It furthers the University's mission by disseminating knowledge in the pursuit of
education, learning and research at the highest international levels of excellence.

www.cambridge.org
Information on this title: www.cambridge.org/9781108075282

© in this compilation Cambridge University Press 2015

This edition first published 1846
This digitally printed version 2015

ISBN 978-1-108-07528-2 Paperback

ILLUSTRATIONS OF

THE VISCOUS THEORY

OF

GLACIER MOTION.

PART I.—CONTAINING EXPERIMENTS ON THE FLOW OF PLASTIC BODIES, AND OBSERVATIONS ON THE PHENOMENA OF LAVA STREAMS.

PART II.—AN ATTEMPT TO ESTABLISH BY OBSERVATION THE PLASTICITY OF GLACIER ICE.

PART III.—ILLUSTRATIONS OF THE VISCOUS THEORY OF GLACIER MOTION.

BY

JAMES D. FORBES, Esq., F.R.S.S. L. AND E.,

Corresponding Member of the Institute of France, and
Professor of Natural Philosophy in the University of Edinburgh.

From the PHILOSOPHICAL TRANSACTIONS.—PART II. FOR 1846.

LONDON:

PRINTED BY R. AND J. E. TAYLOR, RED LION COURT, FLEET STREET.

1846.

XII. *Illustrations of the Viscous Theory of Glacier Motion.*
Part I. *Containing Experiments on the Flow of Plastic Bodies, and Observations on the Phenomena of Lava Streams.*
By JAMES D. FORBES, *Esq., F.R.SS. L. and E., Corresponding Member of the Institute of France, and Professor of Natural Philosophy in the University of Edinburgh.*

Received March 15,—Read April 10, 1845.

§ 1. *Plastic Models.*
§ 2. *Analogy of Glaciers to Lava Streams.*
Note on the Velocity of Lava.

§ 1. *Plastic Models.*

IN the concluding chapter of my " Travels in the Alps of Savoy," I have shown how the obscure relations of the parts of a semifluid or viscous mass in motion (such as I have attempted to prove that the glaciers may be compared to) may be illustrated by experiment.

The larger models, these described and figured, showed very clearly the precise effects of friction upon the motion of such a mass. They were formed of plaster of Paris, mixed with glue, and run in irregular channels, and the relative velocities of the top and bottom, the sides and centre of such a pasty mass were displayed by the alternating layers of two coloured pastes, which were successively poured in at the head of the model valleys. The boundaries of the coloured pastes were squeezed by the mutual pressures into greatly elongated curves whose convexity was in the direction of motion; and in a vertical medial section, the retardation of the bottom and the mutual action of the posterior and anterior parts, shaped the bounding surface of two colours into a spoon-like form.

Now these models convey a very palpable commentary upon the effects of friction on a plastic mass, and likewise on the influence of the mutual pressures of its parts; but in further illustration of the same thing I constructed another model, only executed as the printing of my volume approached its close, and which is cursorily described in a long note (page 377)*, whence its real importance may perhaps have been pretty generally overlooked.

The models in question, of which I have since made many, are formed by accumulating in one end of a long narrow box, AB, Plate IV. fig. 1, a deep pool of the viscid

---

* In this paper reference is of course made to the first edition of my " Travels," the second not having been then published.

material already mentioned, which is retained there by a sluice or partition C which may be withdrawn at pleasure.

The surface of the pool *abcd* is then pretty thickly dusted over with a coloured powder, and the sluice is withdrawn.

The pasty mass subsides slowly under its own weight into the lengthened form *efgh*. The film of colour on the surface is therefore broken up so as to cover three or four times the surface it did at first; and its new distribution marks the lines of greatest separation of the superficial particles of the mass. The appearance of such a model when *run* is shown in fig. 2, and it manifests in the plainest manner the twofold tendency to separation in such a case where the channel is narrow and confined, and there is a certain mass of matter in front. Plate V. shows a more accurate drawing taken from such a model.

The lines of *sliding* separation occur most distinctly marked near the sides, where the friction is greatest, and the central parts are *forced past* the lateral parts, on account of the less embarrassed and consequently swifter motion of the centre; and they incline to the centre although the breadth of the channel be perfectly uniform. But the forces which tear asunder the parts (when such exists) act *perpendicularly to the former* and produce dislocations and fissures, which perfectly correspond to the direction and appearance of the crevasses of a glacier, that is, they are convex upwards or towards the origin of the glacier. It is the former of these lines of separation, or *differential motion,* which constitute and trace out with an exact parallelism the *veined structure* which I have described as forming the normal structure of all true glaciers. Plate V. is a representation of a very beautiful plaster model of more consistence than the other, in which the swelling of the surface and the direction of the open cracks produced by direct thrusts are most beautifully shown; and are even more so in the model than in the engraving. The fissures are transverse and slightly convex to the origin in the higher part of the glacier, then gradually turning round they radiate from a centre in the lower part, exactly as in the glacier of Arolla (Travels in the Alps, Plate VI.), and in all similar cases.

The experiment above detailed was suggested to me by studying the ripple of streams of water, which appear to have the same origin: and in very weak currents moving through very smooth and uniform channels (as the chiseled sides of water conduits) the same may be made manifest by throwing a handful of light powder on the surface, which then becomes divided into threads of particles inclined in the manner I have described at a certain angle from the side towards the centre, depending on the velocity of the stream.

The slightest prominence of any kind in the wall of such a conduit, a bit of wood or tuft of grass is sufficient to produce a well-marked ripple-streak, from the side towards the centre, depending upon the sudden and violent retardation of the lateral streamlets and the freer central ones being momentarily edged away from them. The general course of the motion of the particles is, however, scarcely affected by such a circumstance, for the differential velocities which cause the ripple and the separation, are always small compared to the absolute velocity of the stream; and thus a floating body on the water (just as the moraine on the glacier) perseveres in its course parallel to the side with scarcely any perceptible disturbance. When however the descent is violent and the friction great, floating bodies are gradually drawn towards the centre, and this happens also in exactly the same circumstances to the moraine of the glacier. Plate IV. figs. 3 and 4, shows the relation of the ripple-marks to the channel of a very flat smooth gutter in one of the side streets of Pisa, sketched after heavy rain.

These ripple-marks in water are well seen near the piers of a bridge, or when a post is inserted in a stream and makes a fan-shaped mark in the water cleft by it: such marks have been much neglected by writers on hydraulics; but in one of the most ancient hydraulic treatises, that of LEONARDO DA VINCI, lately printed from the MS. in the Italian collection of writers on hydraulics, they are very well described and figured. A case parallel to the last-mentioned, where a fixed obstacle cleaves a descending stream and leaves its trace in the fan-shaped tail, is well seen in several glaciers, as in that at Ferpêcle, and the Glacier de Lys on the south side of Monte Rosa, particularly the last, where the veined structure follows the law just mentioned. And I desire here to record that the views just presented as to the origin of the veined structure of ice, were confirmed, but were not suggested, by the experiments on viscous fluids just mentioned. The necessity of the tearing up of a solid mass, if it moved at all in a bed presenting insurmountable resistances on all sides, in directions such as the veined structure presents, was foreseen by me whilst dwelling amongst the glaciers themselves, at a distance from books or the means of experiment. The following extract from my Third Letter to Professor JAMESON, written in 1842 from the remote village of Zermatt, contains the substance of all that I have since developed and illustrated at greater length and in different ways rather to meet the difficulties of others, than to confirm what was plainly fixed in my own mind.

"The glacier struggles between a condition of fluidity and rigidity. It cannot obey the law of semi-fluid progression (maximum velocity at the centre, which is no hypothesis in the case of glaciers, but a fact), without a solution of continuity perpendicular to its sides. If two persons hold a sheet of paper so as to be tense, by the four corners, and one move two adjacent corners, whilst the other two remain at rest or move less fast, the tendency will be to tear the paper into shreds parallel to the motion; in the glacier the fissures thus formed are filled with percolated water, which is then frozen. It accords with this view,—1st, that the glacier moves fastest in the centre, and that the loop of the curve described coincides (by observation) with the line of swiftest motion. 2nd. That the bands are least distinct near the centre, for there the difference of velocity of two adjacent stripes parallel to the length of the glacier is nearly nothing; but near the sides, where the retardation is greatest, it is a maximum. 3rd. It accords with direct observation that the *differ-*

ence of velocity of the centre and sides is greater near the lower extremity of the glacier, and that the velocity is more nearly uniform in the higher part; this corresponds to the less elongated forms of the loops in the upper part of the glacier. 4th. In the highest part of such glaciers, as the curves become less bent the structure also vanishes. 5th. In the wide saucer-shaped glaciers which descend from mountain slopes, the velocity being as in shallow rivers nearly uniform across their breadth, no vertical structure is developed. On the other hand, the friction of the base determines an apparent stratification, parallel to the slope down which they fall. 6th. It also follows immediately (assuming it as a fact very probable, but still to be proved, that the deepest part of the glacier moves slower than the surface) that the *frontal dip* of the structural planes of all glaciers diminishes towards their inferior extremity, where it approaches zero, or even inclines outwards, since then the whole pressure of the semi-fluid mass is unsustained by any barrier, and the velocity varies (probably in a rapid progression) with the distance from the soil; whilst nearer the origin of the glacier the frontal dip is great, because the mass of the glacier forms a virtual barrier in advance, and the structure is comparatively indistinct, for the same reason that the transverse structure is indistinct, viz. that the neighbouring horizontal prisms of ice move with nearly a common velocity. 7th. Where two glaciers unite, it is a fact that the structure immediately becomes more developed. This arises from the increased velocity, as well as friction of each due to lateral compression. 8th. The veined structure invariably tends to disappear when a glacier becomes so crevassed as to lose horizontal cohesion, as when it is divided into pyramidal masses. Now this immediately follows from our theory; for as soon as lateral cohesion is destroyed, any determinate inequality of motion ceases; each mass moves singly, and the structure disappears very gradually*."

In explaining the theory of the veined structure at a meeting of the Royal Society of Edinburgh on the 20th of March 1843, I stated that I had arrived at the conclusion that crevasses resulting from tension in certain parts of a glacier, must be formed at right angles to the surfaces of discontinuity or structural veins where they intersect the surface: a law conformable to the empirical one discovered by me on the glacier of the Rhone in 1841†, since generalized in other cases, and which even the adversaries of my theoretical views have admitted to be a correct statement of the facts‡.

My attention was at that time (March 1843) turned by my learned and acute friend Mr. W. A. CADELL, to the veined structure of the slag of iron furnaces as due to the difference of velocity of the parts producing surfaces of separation and peculiar molecular condition. The transition was easy to the case of volcanic rocks and lava

* Third Letter on Glaciers, Edinburgh Philosophical Journal, October 1842.

† Edinburgh Philosophical Journal, January 1842.

‡ Bibliothèque Universelle, tome xliv. p. 153. " C'est en effet un fait assez général que les bandes bleues coupent à angle droit les crevasses," &c.

streams, and this case was pressed on my attention by an unexpected journey which I soon after undertook to Italy and Sicily.

§ 2. *Analogy of Glaciers to Lava Streams.*

There is something pleasing to the imagination in the unexpected analogies presented by a torrent of fiery lava and the icy stream of a glacier. But when we look upon the comparison historically and critically, and find how generally this analogy has been perceived and adverted to by persons of very different views and talents of observation, we are strongly tempted to suspect that some latent cause confers the marked resemblance.

This cause I of course consider to be the laws and condition of their motion, the struggle of a semi-fluid mass of enormous weight creeping down a mountain side, in which fluidity and solidity are so curiously combined, that we should be at a loss in either case how to name it; a straining, crackling, splintering solid, heaved on by the internal energy of the latent fluidity which pervades it, and which at last succeeds in giving to the general character of the motion and the moving mass, those of fluid bodies subject to the law of gravity; whilst the parts, themselves almost rigid, have that rigidity most fantastically subjected to the action of the dominant principle.

In illustration of what has now been said, I shall quote passages from some authors which, without particular research, have come under my notice expressive of the analogy just mentioned.

Mrs. STARKE, the author of a well-known guide book of Italy, published many years ago, speaks of having seen near the crater of Vesuvius in 1818, " five distinct streams of fire issuing from two mouths, and rolling wave after wave slowly down the mountain with the same noise (?) and in the same manner *as the melting glaciers roll into the valley of Chamouni*; indeed this awful and extraordinary scene would have brought to mind the Montanvert, had it not been for the crimson glare and excessive heat of the surrounding scoriæ*."

Mr. AULDJO, an intrepid alpine traveller, writing about Vesuvius, in 1832, says, " The field of lava in the interior of the crater, inclosed within a lofty and irregular bank, might be likened to a lake whose agitated waves had been suddenly petrified; and in many respects resembles the *Mers de Glace*, or level glaciers of Switzerland, although in its origin and materials so very different†." And the view in the same work of " streams of lava on the south-east of the cone" presents a perfect analogy to a glacier, bearing on its surface three medial and two lateral moraines.

Captain BASIL HALL, writing of Vesuvius at a later period, uses these remarkable expressions whilst describing an eruption of lava :—" The colour of this stream was a brilliant pink, much brighter at the sides than in the middle, where either from the cooling of the surface, or the accumulation of cinders and broken pieces of stone, a

* STARKE's Traveller's Guide, Ninth Edition, p. 293.
† AULDJO's Sketches of Vesuvius, p. 10, published 1833.

sort of dark ridge or backbone was visible from end to end, not unlike the moraine on the top of a glacier. This reminds me of a curious analogy which often struck me, between two objects so dissimilar as a glacier and a lava stream. They are both, more or less, frozen rivers; they both obey the law of gravitation with great reluctance, being essentially so sluggish, that although they both move along the bottoms of valleys with a force well nigh irresistible, their motion is sometimes scarcely perceptible*." This remarkable passage, worded with the usual scrupulous care of the author, combined with his account of the mechanism of a glacier in the description of the glacier of Miage in the same work, show that he had arrived at more correct notions on the subject than any of his contemporaries; notions which chiefly required careful observation to give them the force of demonstration. The allusion to *moraines* as characteristic of lava streams as well as glaciers, in the preceding extract, is perfectly borne out by the view of the lava of 1831 given by Mr. AULDJO and already cited; the same appearance is mentioned by M. ELIE DE BEAUMONT in his account of Etna in the following terms: "Une des circonstances que les coulées de lava présentent le plus invariablement * * * * consiste en ce que chaque coulée est flanquée de part et d'autre par une digue de scories accumulées qui rapellent par sa forme la moraine d'un glacier, * * * souvent aussi les coulées présentent de pareilles digues vers leur milieu, lorsqu'elles sont partagées en plusieurs courants distincts coulant l'un à coté de l'autre†."

In another place the same author compares the movement of the upper crust of the lava to that of glaciers according to the then prevalent theory:—"L'écorce supérieure d'une coulée separée de l'écorce inférieure et du sol sousjacent par une certaine épaisseur de lave liquide, ou du moins visqueuse, se trouve dans un état comparable à celui d'un glacier, qui, ne pouvant adhérer au sol sousjacent à cause de la fusion continuelle de sa couche inférieure, se trouve contraint de glisser‡."

Finally, M. RENDU, Bishop of Annecy, in his excellent Essay on Glaciers, refers in one passage (and I believe in one only) to the possible analogy with a lava stream, " [le glacier] s'affaisse-t-il sur lui-même pour couler le long des pentes comme le ferait une lave à la fois ductile et liquide§?"

The following considerations seem to show more than a general external analogy between lava streams and glaciers.

Their velocities are sometimes equally slow. Although common lava is nearly as liquid as melted iron, when it issues from the orifice of the crater, its fluidity rapidly diminishes, and as it becomes more and more burdened by the consolidated slag through which it has to force its way, its velocity of motion diminishes in an almost

---

* Patchwork, by Captain HALL, vol. iii. p. 118, published 1841.

† Recherches sur le Mont Etna, p. 184. Published in the Mémoires pour servir à une Déscription Géologique de la France, tome iv. 1838.

‡ Ibid. p. 177.

§ Théorie des Glaciers, Mém. de l'Académie de Savoie, tome x. p. 93, published 1841.

inconceivable degree, and at length, when it ceases to present the slightest external trace of fluidity, its movement can only be ascertained by careful and repeated observations, just as in the case of a glacier.   In November 1843, I watched lava issuing rapidly from a small mouth in the crater of Vesuvius at the rate of about one foot in a second.   The eruption of Etna in 1832 advanced at the rate of five miles in two days, which is at the rate of one foot in about six seconds*.   We may contrast with this the eruption of Etna in 1614, which yielded a lava which advanced but two miles in *ten* years according to DOLOMIEU †, during the whole of which time its motion was sensible.   This gives a mean rate of rather more than three feet per day ; but at the conclusion it was no doubt much slower.

Mr. SCROPE ‡ saw the lava of 1819 in the Val del Bove moving down a considerable slope at the rate of a yard a day, nine months after its eruption.   It had, he adds, the appearance of a huge heap of rough cinders ; its progression was marked by a crackling noise due to friction and straining, and " on the whole was fitted to produce *any other idea than that of fluidity.   In fact,*" he continues, " we must represent to ourselves the mode in which the crystalline particles of lava move amongst one another, rather as a sliding or slipping of their plane surfaces over each other, facilitated by the intervention of the elastic (?) fluid, than as the rotatory movement which actuates the molecules of most other liquids."   It is generally conformable to this view that we find in HAMILTON's *Campi Phlegræi* (fol. 1. 38. *Note*) the curious remark that some lava is so incoherent, or whilst fluid has so little *viscosity,* that in issuing from the volcano (Vesuvius) it has appeared "*farinaceous,* the particles separating as they forced their way out, just like meal coming from under the grindstones."

From all this it is quite clear that the seeming rapidity of the parts of a glacier, or the slowness of its motion, cannot be taken as the slightest evidence of its moving otherwise than as a fluid, contending with the *rigor* of the parts which include and resist the moving force, which is truly hydrostatic though limited in its exercise.

It is manifestly futile and unphilosophical to seek one *cause* of motion in a lava which, like that of Vesuvius in 1805, must have described as many hundred feet in a *minute* as that of 1614 from Etna probably did in a *year* § ; for the *mean* daily motion of the latter during *ten years* was three feet ; but toward the end of that time it must evidently have had for a long period an average motion of one-half or one-quarter of this, and therefore below the observed mean movements of certain glaciers. Fluidity, in the first instance as in the second, was the propelling vehicle or manner in

* E. DE BEAUMONT, Recherches sur le Mont Etna.

† Quoted by E. DE BEAUMONT, p. 85.   The original is in the Journal de Physique, vol. i. of the New Series, where it is mentioned that the same slowness of motion has been observed in lavas of Vesuvius.   FERRARA (Descrizione del Etna.   Palermo, 1818) denies this statement, but not I think on sufficient grounds.

‡ On Volcanoes, p. 102.

§ See Note on the Velocity of Lava Streams at the end of this paper.

which gravity acted, and this is a sufficient answer to any attempt to maintain that the plasticity of a glacier is a collateral but not a primary cause of motion,—a distinction surely without a difference.

As in the case of all imperfect fluids, the central and superficial particles move faster than the lateral and inferior ones ; and when the fluidity is *exceedingly* imperfect, as in those long-flowing lavas, there must be a rupture of continuity between the parts to permit them to slide and jostle past one another.   This is evidently the cause of the noise referred to by Mr. SCROPE and other writers.   This tearing up of the stream into longitudinal stripes, occasioned by the varying velocity of the parts, is thus described by M. DUFRENOY in his account of Vesuvius : " La plupart des coulées présentent des bandes longitudinales assez parallèles entre elles : ces larges stries saillantes sur la surface sont les traces du mouvement de la lave qui ne s'avance pas d'une seule pièce, mais par bandes parallèles*."

And M. ELIE DE BEAUMONT describes a lava stream at Etna in these terms : " La surface offrait de profondes cannelures parallèles entre elles, dirigées dans le sens du mouvement qui l'avoit déversée à l'extérieur et qui étaient croisées par *de nombreuses gerçures transversales*†."   Here then is evidently the twofold system of rents and perpendicular fissures described in the commencement of this paper as being found in the models, and as being conformable to the phenomena of glaciers.

During the winter 1843–44 which I spent in Italy, I had an opportunity of testing these resemblances, and tracing others to glaciers in the lavas of Vesuvius and Etna. I entered on the inquiry with a very jealous care of being drawn into the admission of fanciful or imperfect analogies ; and I shall confine myself to the statement of one or two most plain and undeniable confirmations, selected from the results of many fatiguing rambles.

The plastic nature of the viscous lavas of Vesuvius and Etna is such as well might obliterate any internal traces of rents due to differential velocity, which, in the mass, are speedily closed and reunited as in a stream of treacle, or in the plaster models before explained, where the interior is homogeneous and the superficial coating above is permanently dislocated.

In lavas the indescribable ruggedness of the surface very generally prevents any record of the gentler play of forces.   The following facts appear to me quite conclusive as to the manner in which a mass partially solidified, yet moving as a fluid, is torn up by the interior forces which act upon it.

1. At Vesuvius, the *Fossa della Vetrana* between the Hermitage and Monte Somma, is a valley lined with the lava of 1751.   I here observed that the lava was in some places detached from the wall of the valley, leaving a cavity on the sheltered side of a projecting elbow of rock, just as a glacier does in similar circumstances, showing the considerable consistence which the lava possessed.

* DUFRENOY sur les Environs des Naples, p. 324.
† E. DE BEAUMONT, p. 38.

In the upper part of this Fossa the lava has a distinct linear structure where broken, in shells parallel to the sides, whose thickness varies from one-third of an inch upwards. The position of these surfaces of dislocation is indicated (for illustration) in figure 5 of Plate IV.

2. In the vast lava wastes of Etna, we encounter not only a greater extent of surface, but a greater variety of condition as to cohesion of the lava streams, and the slope down which it has descended, and thus we have a better chance of meeting with specimens of the manner in which the semi-solid crust of a lava stream is torn up and crevassed by the effect of gravity compelling it into the circumstances of fluid motion. From this tendency of all lavas to form slags, and of these slags to be splintered, tossed, and remoulded by the action of the still liquid portion of the stream below or around, not one-thousandth of the surface bears marks of the simple condition of fluidity under which it was originally moulded; and though when viewed from a distance, and in connexion with the form of the ground over which it has passed, we see plainly enough that it has *flowed* like a stream, the absence of any trace of easy undulating forms which characterise fluids or plastic masses, give to the *sciarre* of Etna (the *cheires* of Auvergne) an appearance far more removed from pristine fluidity than the glacier masses of Switzerland.

In traversing many miles of lava wastes between Nicolosi and Zafarana, on the eastern slope of Etna, I met with one singularly favourable specimen of a branch of a stream consolidated exactly as it had moved, and undisturbed afterwards. It is the part of the current of 1763, called *Lava delle Cerve*. The branch stream in question may be ten yards wide, and presents a thin crust, which has floated on the viscid lava below, and which, while yet imperfectly solidified, has been urged to move with the rest of the stream, and has undergone a process of division and rending accordingly. The stream has flowed in the direction from left to right in figure 6. The lateral parts PP, QQ have been *literally torn to pieces longitudinally* (as I wrote on my note-book on the spot) by the multiplied rents which showed the dislocation of the quicker moving central from the lateral parts, and these rents *inclined towards the centre of the stream in the direction in which it moved.* The length of the stream was divided by transverse rents strikingly convex towards the origin of the stream, as shown in the same figure. These cracks were marked by another peculiarity; the cake of floating scoria had not only been cracked across but pushed *upwards*, generally *forwards and upwards*, before it was finally included in the cooling mass of the stream; the result was the arrangement shown in the longitudinal section, fig. 7, which it will be seen resembles the tiling of a house, only that the fractured parts do not always overlap, but the anterior edge is tilted upwards. It will thus be seen that this tendency to separation acts also in the *vertical* plane, and the dotted lines *aa'*, *bb'*, &c. indicate the direction of its action, coinciding with the surfaces of differential motion, which produce what I have called the *frontal dip* of the veined structure of the ice of glaciers.

3. At no great distance from this lava, and near the foot of the hillock called the Serra Pizzuta, between the last-named point and the valley of Tripodo, I observed a *transverse* section of a lava stream exhibiting an arrangement in bands or plates, nearly parallel to the side of the current, but inclining towards the centre.

4. Between Zafarana and the Porta Calanna (Etna), a remarkably pretty illustration occurs in the surface of an old lava stream, worn and polished by the action of a brook. Where the lava has had to turn an abrupt corner of a rock, A, figure 8 (which represents a ground plan), the progress of the lava being violently checked by the resistance of the projecting mass, has been torn up into longitudinal shreds, which from imperfect fluidity have not reunited, but have left open cavities of the form represented in the figure, which exhibit with remarkable fidelity the forms of the fissures with which glaciers are sometimes traversed, when they are subjected to sudden transitions in their states of motion (as in the glacier des Bossons at Chamouni), and which coincide in direction with the veined structure, and pass into it by imperceptible gradations.

5. What I have called the *frontal dip* of the veined structure in glaciers*, I have explained by the accumulation of a sluggish mass of considerable extent upon a floor or bed offering the resistance of intense friction; in consequence of which the mass of ice, urged downwards and forwards by its intense weight, being resisted by the friction of that which immediately precedes it, must yield in the direction of least resistance, or squeeze itself in a slanting direction forwards and *upwards*, and thus sliding over the resisting mass immediately in front, will produce surfaces of discontinuity or differential velocity in that direction. Such a result I inferred from general principles without reference to any particular example, and the explanation of the superficial convexity of the lower part of many glaciers was evidently satisfactorily explained by it.

The convex swelling form of a viscous stream will depend principally upon the relative measure of two quantities, the stiffness or viscosity of the fluid, and the inclination of the surface; although it will also depend on the part of the stream, whether near the origin or the termination, which we consider.

I have found this variation from concave to convex, depending upon circumstances, alike in glaciers and lava streams. Some very highly inclined small glaciers existing at considerable heights, and therefore very hard and consistent, are, nevertheless, deeply concave from end to end, the slope compensating for the stiffness of the matter; such is a beautiful glacier, named, as far as I can learn, La Gria, or Glacier de Bourget, which descends from the Aiguille de Gouté towards the valley of Chamouni. See Plate IV. fig. 9.

Many, perhaps most, lava streams, where they have well-determined banks, are concave during the longer part of their course, but towards their termination they

---

* See my Travels in the Alps, 1st Edit. pp. 167, 376, and letter to Dr. WHEWELL in JAMESON's Journal, Oct. 1844.

become convex as their viscosity increases.   Nevertheless, I have seen portions of well-bounded streams decidedly convex.

The appearance of the termination of a lava stream approaches strikingly that of a glacier.   But this is much more than a vague analogy, and the accounts of faithful eye-witnesses prove the resisted motion of the doughy stream to be such as I antici-pated.   We find it explicitly stated over and over again in the writings of DOLOMIEU* and DELLA TORRE † (and more particularly by the latter), that when a lava stream meets with any obstacle in front which checks its course, or when its course is checked by its own sluggishness, the stream swells, and gains gradually in thickness by the fluid pressure from behind urging its particles *forwards* and *upwards*.   So striking was this natural effect of semi-fluid pressure, that these old observers attri-buted it to a peculiar force developed in the lava, of the nature of " fermentation," producing intumescence, the only way by which they could account for the vertical rise of the fluid, although it was very evident that the result was only what might be expected from the nature of the lava.   It was also observed that when the lava stream had thus attained a certain height, it began to move on again, the necessary result of the increased hydrostatic pressure, although attributed by the authors named to the heat developed by chemical action.   The tenacity with which the idea was long adhered to, that the residual fluidity of a nearly cooled lava stream was insufficient to account for its progress, without attributing to it the qualities of a second volcanic focus, are curious proofs of how long a palpable cause may be rejected as insufficient to explain a phenomenon, and a totally imaginary one superadded ‡.

I may add, that lava streams sometimes push their extremities *up hill* §; glaciers do the same.

In addition to the considerations already stated, which illustrate the viscous theory of glaciers, I am glad to avail myself of two which have reached me from indepen-dent and impartial sources.

The first is by Mr. DARWIN, who in a small book on " Volcanic Islands," published about the time that I was engaged in making the preceding observations on Etna and Vesuvius, pointed out in a very clear manner the explanation which the veined struc-ture of glaciers lends to that of volcanic rocks belonging to the Trachytic and Obsidian Series, where the lamination, instead of being obscure and rare, as it generally is in the Augitic lavas, owing perhaps to their greater fluidity, and more viscid and homogeneous texture, is the general rule.   " The most probable expla-nation," says Mr. DARWIN, " of the laminated structure of these felspathic rocks

---

* Papers in the Journal de Physique.

† Histoire du Vésuve. Naples, 1771, 8vo, p. 207–9, and several other places.

‡ See the view of the termination of a lava stream in Auldjo's Sketches of Vesuvius, facing p. 92.   The reader may also compare the view of a grotto in the lava, in the same work, with that of the source of the Arveiron, in my Travels, p. 387.

§ HAMILTON, Campi Phlegræi, folio, vol. i. p. 40, *note.*

appears to be that they have been stretched whilst flowing slowly onwards in a pasty condition, in precisely the same manner as Professor Forbes believes that the ice of moving glaciers is stretched and fissured. In both cases the zones may be compared to the finest agates; in both they extend in the direction in which the mass has flowed, and those exposed on the surface are generally vertical*."

The other illustration is contained in a communication with which I have been favoured by Mr. Gordon, Professor of Civil Engineering in Glasgow, and which has been printed in the Philosophical Magazine for March 1845, to which therefore I may refer. I need only state at present that it demonstrates, from observations on the flow of Stockholm pitch with a speed wholly insensible, and which requires some months for its accomplishment even in small masses, that a motion, of the nature of fluid motion, takes place at temperatures at which the pitch remains so hard as to be fragile throughout, and presents angular fragments with a conchoidal fracture. Mr. Gordon adds, that the resistance of the pitch to its own forward motion produces bands of differential velocity and having the *frontal dip*.

Edinburgh, February 26, 1845.

Note on the Velocity of Lava, referred to in p. 149.

The following are a few facts which I have collected on the velocity of lava. That of Vesuvius in 1805 appears to be the most fluid on record. Von Buch, who was in company with MM. de Humboldt and Gay-Lussac, describes it as shooting suddenly before their eyes from top to bottom of the *cone* in one single instant†, which must correspond to a velocity of many hundred feet in a few seconds without interpreting it literally. Melogrami, quoted by Breislak‡, says it described three miles in four minutes, or about seventy-five feet per second *at a mean*. The same lava, when it reached the level road at Torre del Greco, moved at the rate of only eighteen inches per minute, or three-tenths of an inch per second§. The lava of 1794 (Vesuvius) reached the sea, a distance of 12,961 feet, in six hours, or passed over one-third of a mile per hour, or eight inches per second‖; whilst the lava of Etna, in 1651, described sixteen miles in twenty-four hours, or above a foot per second the whole way. That of 1669 (Etna), which destroyed Catania, described the first thirteen miles of its course in twenty days, or at the rate of 162 feet per hour, but required twenty-three days for the last two miles, giving a velocity of twenty-two feet per hour¶; and we learn from Dolo-

---

* Darwin on Volcanic Islands, 1844. The whole passage, pp. 65–72, illustrates this analogy.

† Bibliothèque Britanique, vol. xxx. The vertical height of the cone proper is 700 or 800 feet; the length of the slope may therefore be 1300 feet.

‡ Institutions Géologiques, iii. 142.

§ Nicholson's Journal, vol. xii.   ‖ Breislak, Campanie, i. 203.

¶ Ferrara, Descr. del Etna, p. 105. This appears from the dates, though at variance with one assertion of the author.

MIEU, that this same stream moved during part of its course at the rate of 1500 feet an hour, and in others took several days to cover a few yards*.

The lava of 1753 (Vesuvius), starting with a velocity of 2500 feet per hour, soon diminished to sixty feet†, as did that of 1754 to the same‡; and of 1766 to thirty feet per hour§. The lava of 1831 (Vesuvius) moved over 3600 feet in twenty-six hours, and finally advanced steadily at the rate of ten feet an hour‖. The lava of Etna of November 1843, is said to have moved over three paces per second at the distance of a mile from the crater.

The stream of 1761 (Vesuvius), before it stopped flowing, advanced but three yards a day¶; and that of 1766, which continued moving for about nine months, moved over but a small space in that time. Had the attention of authors been equally directed to the *slow* as to the rapid advancement of lava, there is no doubt that we should find many instances besides these recorded by DOLOMIEU and SCROPE, of continuous movements of three feet, and even one foot a day, or less.

---

* DOLOMIEU Isles Ponces, p. 286. Note.   † DELLA TORRE, Histoire, &c., p. 196.

‡ Ibid. p. 130.   § HAMILTON, Campi Phlegræi, i. 19.

‖ AULDJO, Sketches of Vesuvius, p. 79, with a sketch of the front of the stream whilst advancing at this rate.

¶ DELLA TORRE, p. 182.

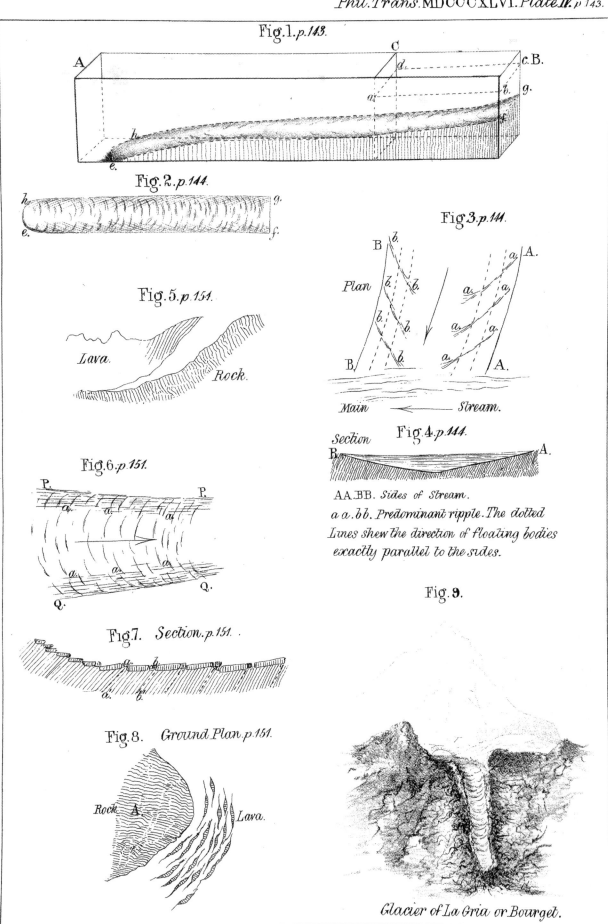

Fig. 1. p. 143.

Fig. 2. p. 144.

Fig 3. p. 144.

Plan

Main ← Stream.

Fig. 5. p. 154.

Lava.

Rock.

Fig. 4. p. 144.

Section

AA. BB. *Sides of Stream.*
a a. b b. Predominant ripple. The dotted Lines shew the direction of floating bodies exactly parallel to the sides.

Fig. 6. p. 151.

Fig. 9.

Fig. 7. Section. p. 151.

Fig. 8. Ground Plan. p. 151.

Rock A Lava.

Glacier of La Gria or Bourget.

J. Basire, Lith.

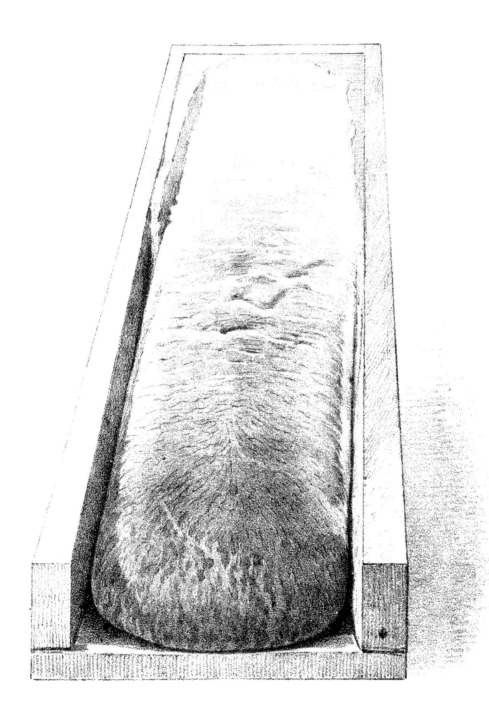

J. Basire Lith

XIII. *Illustrations of the Viscous Theory of Glacier Motion.*

Part II. *An attempt to establish by observation the Plasticity of Glacier Ice.*

By James D. Forbes, *Esq., F.R.S.S. L. and E., Corresponding Member of the Institute of France, and Professor of Natural Philosophy in the University of Edinburgh.*

Received July 28, 1845,—Read January 15, 1846.

§ 3. De Saussure's *Theory.*

§ 4. *Modifications of* De Saussure's *Theory.*

§ 5. *Experiments at Chamouni on the Plasticity of Ice.*

§ 3. De Saussure's *Theory.*

WHEN Gruner proposed the explanation of glacier motion by the sliding of the ice over its bed, and De Saussure illustrated and confirmed it by considerations drawn from the lubricating action of the earth's heat melting the ice in contact with the soil*, there is no reason to suppose that either of them thought it necessary to take into account the varying form of the channel through which the glacier had to pass, and the consequently invincible barrier presented to the passage of a rigid cake of ice through a strait or narrow aperture when it occurred. This is the more remarkable, because he conceives that the *inequalities of the bed or bottom* may be overcome by the hydrostatic pressure of the water, which he supposes may be imprisoned between the rock and the ice, so as absolutely to heave the latter over the resisting obstacles.

I believe that in no part of De Saussure's writings will there be found any, the slightest reference to the possibility of the glacier when fairly formed *moulding* itself to the inequalities of the surfaces over which gravity urges it; nor is there any trace of the correlative fact of an unequal motion of the sides and centre of the ice, which may in some sense be considered as the geometrical statement of the preceding physical fact. The fact of plasticity was suspected by Basil Hall, and more distinctly announced by Rendu, as shown in the first part of this paper; but it could not be proved until the geometrical fact of the swifter motion of the centre of the glacier relatively to the sides was established in 1842†. The contrary opinion at that time

---

* To do Gruner justice, he appears to have been aware of the effects of the earth's heat and the lubricating action of the water thawed from the glacier : " Lorsque les côtés de l'amas [de glace] qui touchent la montagne, fondent en entier, toute la masse entraînée par son poids glisse sur son fond et s'avance dans la vallée," French translation, p. 333 ... " il est vraisemblable que leur surface inférieure [*i. e.* des glaciers] se liquéfie autant, et peut-être plus que la supérieure," ib. p. 289.

† Edinburgh Philosophical Journal, October 1842, and Travels in the Alps of Savoy, p. 134.

generally entertained would have been conclusive against the hypothesis of plasticity called forth by the gravity of the mass.

So far, then, as appears from his writings, De Saussure considered the ice of glaciers to constitute a mass possessing rigidity in the highest degree, such rigidity in short as common experience assigns to ice tranquilly frozen in small masses, which is sensibly inflexible. It is in this sense in which I have spoken of De Saussure's sliding theory, as one which " supposes the mass of the glacier to be a rigid one sliding over its trough or bed in the manner of solid bodies*," and I adhere to the definition as excluding the introduction of the smallest flexibility or plasticity, to which the term rigidity is correctly opposed. I consider too that De Saussure's theory supposes the mass of the glacier to slide over its trough or bed in the manner of *solid* bodies, that is, not as a heap of rubbish or absolute fragments, such as a glacier sometimes precipitates over a rock, but which evidently did not enter into De Saussure's explanation, nor, in fact, required any theory.

As to the crevasses which form so prominent a feature of many glaciers (although many are in parts almost devoid of them), I do not recollect that De Saussure alludes to them as *facilitating* in any way the movement of the glacier, but simply as *results* of its motion and of the rigid character of ice. And I believe that this view (whether it was held by De Saussure or not) is substantially correct. The regular system of crevasses of a glacier is approximately transverse, rather arched upwards towards the origin of the glacier, and as De Saussure supposes the glacier to be pressed downwards by the mass of snow accumulating at its head, it is hard to believe that he could have regarded these fissures as in any way essential to its movement, even were they very numerous ; the tendency of such a pressure from above would rather seem to be to pack the ice like an arch, opposing its convex side to the direction of the pressure.

The view now given of De Saussure's theory of glacier motion is not only conformable to what may be gathered from his writings, but expresses the unanimous understanding of his numerous commentators, followers and opponents. As some doubt has lately been hinted as to the definiteness of De Saussure's conception of a glacier as a mass devoid of flexibility and plasticity and urged down a slope *as a whole*, by the lubricating action of fusion in contact with the soil to an extent which, in extreme cases, might even give it the character of buoyancy, I will take the liberty of quoting some indisputable authorities amongst writers of name in different countries.

And first from De Saussure himself :—

" La chaleur de la terre fait fondre les neiges et les glaces, même pendant les froids les plus rigoureux lorsque leur épaisseur est assez grande pour préserver du froid exterieur les fonds sur lesquels elles réposent."—*Voyages*, § 532. *** " C'est elle qui entretient les torrents, qui, même pendant les plus grands froids, ne discontinuent jamais de sortir de tous les grands glaciers." § 533. ***

* Travels, p. 362.

"Presque tous les glaciers reposent sur des fonds inclinés ; et tous ceux d'une grandeur un peu considérable ont au-dessous d'eux, même en hiver, des courans d'eau qui coulent entre la glace et le fond qui la porte. On comprend donc que ces masses glacées, entraînée par la pente du fond sur lequel elles reposent, dégagées par les eaux de la liaison qu'elles pourraient contracter avec ce même fond, soulevées même quelquefois par ces eaux, doivent peu à peu glisser et déscendre en suivant la pente des vallées, ou des croupes qu'elles couvrent." § 535.

"Quand on considère que ces glaces reposent sur des plans inclinés, qu'il coule sous elles des torrens d'eaux qui les fondent par en bas, les détachent et les soulevent, ne sent-on pas que leur permanence dans la même place est une chose *physiquement impossible* ?" § 2284.

RAMOND's account :

"La cause [de la marche des glaciers] est facile à concevoir : une masse qui pèse sur un plan incliné tend nécessairement à déscendre, et cette tendance est favorisée dans les glaciers par le choc des torrents qui roulent sous leur voûtes, par l'humidité que leur masse communique au terrain qui les porte, enfin par cette multitude innombrable de cavités qui creusent leur partie inférieure, et dont l'effet est de diminuer le frottement en diminuant l'étendue des surfaces *."

ELIE DE BEAUMONT's account of DE SAUSSURE's theory: speaking of the lava of Etna, he says, "L'écorce supérieure d'une coulée separée de l'écorce inférieure et du sol sousjacent par une certain épaisseur de lave liquide, ou du moins visqueuse, se trouve dans un état comparable à celui d'un glacier, qui, ne pouvant adhérer au sol sousjacent à cause de la fusion continuelle de la couche inférieure se trouve contraint à glisser †."

BISCHOFF's account :

"Das jährliche Vorrücken der Gletscher welches SAUSSURE ganz einfach aus einem allmäligen Herunterrutschen der unteren durch das Aufthauen des Eises schlüpfrig gewordenen Seite des Gletschers auf der schiefen Fläche des Bodens erklärt, ist eine bekannte Thatsache ‡."

AGASSIZ's account :

"Autrefois on admettait tout simplement qu'ils glissaient sur leur fond, en vertu de leur propre pesanteur, et que ce glissement était favorisé par les eaux au fond de leur lit. C'était l'opinion de SAUSSURE §."

MARTINS' account :

"DE SAUSSURE, ESCHER DE LA LINTH, ANDRÉ DE LUC, attribuent cette progression au poids des glaces et à l'affaissement produit par la fonte de la face inférieure qui repose sur le sol ‖."

---

* Voyages en Suisse par COXE, traduit par RAMOND, ii. 119, 1790.

† Mémoires pour servir à la Déscription Géologique de la France, iv. 177, published 1838.

‡ Wärmelehre, p. 180, published 1837.   § Etudes sur les Glaciers, p. 152, published 1840.

‖ Sur les Glaciers de Spitzberg. Bibliothèque Universelle, 1840, tom. xxviii. p. 166.

STUDER's account:

"Die bisher fast allgemein herrschende Theorie erklärte die Bewegung der Gletscher aus der Schwere allein. Es soll die Gletschermasse als STARRER KÖRPER auf ihrer Felsgrundlage, wie auf einer schiefen Ebene, theils durch ihr eigenes Gewicht theils durch dem Druck der höheren Eis- und Firnmasse herunter gleiten (GRUNER, RAMOND, KUHN, DE SAUSSURE, ESCHER)*."

This last testimony of the most exact and most learned of the living Swiss geologists as to the sense in which DE SAUSSURE's theory has always been understood is so important that I shall add a translation: "The hitherto generally prevailing theory explains the movement of the glaciers by gravity alone. The glacier masses are considered as RIGID BODIES, which slide down over their rocky beds partly by their own weight, partly under the pressure of the higher ice and *névé*." My interpretation of the views of DE SAUSSURE as regards the rigidity of the glacier ice is thus borne out by an independent authority, for M. STUDER's work and my own appeared simultaneously. It is further confirmed by private communication with another eminent Swiss naturalist nearly connected by relationship with DE SAUSSURE himself, who is more intimately acquainted with the opinions and writings of his illustrious kinsman than any other person now alive, and who considers that DE SAUSSURE's views were confined to the general analogy of the glaciers to solid masses sliding down inclined planes, and that the effects of the inequalities of the channels and forms of the ice-basins were not comprehended in his theory.

If we feel surprise that a naturalist and observer so eminent had not adverted to the difficulty of imagining a solid cake of ice, even though perfectly detached from its bed, to disengage itself from the obstacles and sinuosities of its rocky channel, we should remember,—*first*, that the explanation is given in the most general terms, and there is no appearance that its author looked more closely at its consequences and details than to satisfy himself that a sliding motion in the abstract was rendered possible by the action of the earth's proper heat, an ingenious and philosophical element of the theory (however inadequate), and that which being due principally to DE SAUSSURE, renders the theory properly his, and connected it with his ingenious inquiry into this curious part of physics as a distinct and wholly independent investigation. *Secondly.* Every one knows how an application of a principle so true and so ingenious leads men of even the most exact habits of thought to overlook difficulties in a subject almost unstudied. DE SAUSSURE did much for our knowledge of glaciers, and he saw much which no one had observed before him: we must not blame him if, yielding to a true and natural analogy of sliding bodies, he overlooked real and great difficulties inherent in the conception of a glacier as a solid continuous mass and highly rigid. *Thirdly.* In DE SAUSSURE's time no plan or map, worthy of the name, of any glacier existed, and this was a blank which even DE SAUSSURE did not attempt to supply. The popular notion of a glacier, which it is certain he had in his mind

* Lehrbuch der physikalischen Geographie und Geologie, 1844.

when he penned the passages which relate to their motion, is a mass of ice of small depth and considerable but uniform breadth sliding down a uniform valley, or pouring from a narrow valley into a wider one, as is the case with a vast majority of glaciers tolerably accessible, and which alone were visited at the time of publication of the first edition of the *Voyages dans les Alpes*. In all these cases the lateral resistance might easily be overlooked, and the popular comparison to one solid body sliding on another and lubricated by its own liquefaction might be accepted as a complete explanation; as has even been done at a later period by those who have attempted to illustrate DE SAUSSURE's theory by experiment, but who, like him, neglected the form and undulations of the bed in which it rests.

§ 4. *Modifications of* DE SAUSSURE's *Theory.*

DE SAUSSURE and his immediate followers appear to have considered the crevasses which occur transversely in most glaciers, as the result of the inequalities of the beds down which they are constrained to move; but other writers have imagined that the part which these crevasses perform in the phenomena of glacier motion is fundamental, and essential to the existence of the movement at all. Some writers have remarked that the fall of ice blocks over the precipice which often occurs near the lower end of glaciers, leaving the superior portions unsupported, allows them to advance to fill the position formerly occupied by the portion of the now fallen ice. But in this case it would appear that cause and effect are in some degree confounded. The ice about to be projected over the cliff must either advance towards its fall by its own gravity, or by the pressure of the parts behind. If its own gravity suffices, the same cause will urge the ice behind it to move similarly, whether the block in question fall or not; and if it be the pressure from behind which shoves it on, then still more is the pressure of the entire glacier the cause of motion of the entire glacier, irrespective of the precipitation of its more advanced part.

Thus, M. MARTINS' theory of the progression of glaciers is, that the weight of the parts causes them to separate by fissures into wedge-shaped masses, without their sliding along the bottom; that the fissures become filled with frozen snow, and that thus the glacier is perpetuated and extended year by year. "Cette progression," he says, " n'est donc ni un glissement ni un affaissement difficiles à comprendre, puisque la glace doit adhérer au sol, mais un démembrement successif*." Besides other objections, it is now universally admitted that the glacier-proper does not grow by the consolidation of snow in its fissures.

But setting aside the attempt to render the sliding motion of the entire glacier considered as a plane slab more easy, by considering the motions of the parts instead of the motion of the whole, we are led to notice the attempt to reconcile the sliding theory to recent observation, by ascribing to the crevasses of the glacier the important office of enabling it to accommodate itself to the inequalities of its channel.

* MARTINS sur les Glaciers de Spitzberg et de la Suisse. Bibl. Univ. Juillet 1840.

Our object is, in this section, merely to state the view in its most plausible form, which in the succeeding section we shall controvert by experiments giving it a direct negative. In the third portion of this essay we shall enter more at large into the phenomena of crevasses, and mention other objections to this hypothesis and every modification of it.

According to this view, the friction of the ice against the sides of the valleys will produce a dislocation of the glacier into longitudinal stripes (as shown in Plate VII. fig. 1.*), where a transverse line bb' becomes by the irregular motions of the ice distorted into the zigzag form $hcc'h$. Or if we suppose the plasticity of the ice to be sensible, but that its action is accompanied with fractures, the abruptness of the angles of the figure will be softened, as in the broken line $lmm'l$ in the lower part of the same figure. This latter hypothesis evidently merges into the true plastic theory, when the part of the progression due to the flexure of the transverse lines bears a large proportion to the effect of the longitudinal slide, or more generally, when the surfaces of sliding or yielding become greatly multiplied, when the notched line will merge into a curve.

The passage of the glacier through a gorge or contraction is explained on the same view by figure 2, where the resistance of the sides having occasioned a series of parallel longitudinal rents as before, the portion of the glacier beyond the limits of breadth of the gorge BB' is supposed to be detained or embayed whilst the intermediate columns slip through.

§ 5. *Experiments at Chamouni on the Plasticity of Ice.*

It has been shown that in order to reconcile DE SAUSSURE's theory of sliding motion with the ascertained fact that the centre of the glacier moves faster than the sides, it had been assumed that solutions of continuity or longitudinal crevasses were formed parallel to the length of the glacier, by means of which the central portion slides past that adjacent to it, and so on for successive strips as we approach the sides, the more rapid retardation near the sides being rendered mechanically possible by the increased number of these longitudinal dislocations.

The result was therefore predicted to be that the glacier would be found to move by *echelons,* or that strips of ice of a certain number of feet, or yards, or fathoms, would move either suddenly or by gradual sliding, but at all events so as to mark by an abrupt separation at the longitudinal fissure, that the one portion of ice has slipped past the other by a distinct measurable quantity.

When I first learnt at Geneva, in August 1844, from Mr. HOPKINS's published papers†, that this was really the author's meaning, it occurred to me that the proof

---

* These figures and their interpretation are taken from Mr. HOPKINS's First Memoir in the Cambridge Transactions, vol. viii. part 1. A figure similar to the first is to be found in a more recent paper by the same author in the Philosophical Magazine for June 1845.

† Cambridge Transactions, vol. viii.

between the rival theories was easy, and that it was only necessary to place a series of marks in a right line transversely to the glacier, and observe whether they were displaced by an imperceptible flexure, or whether they slid past one another by sudden dislocations.

Such a proof was independent of any assertion as to the existence or not of such fissures as those contended for, about which different opinions might be formed, especially as they might be asserted to exist although invisible to the eye. Being satisfied in my own mind of the non-existence of such fissures wherever the ice is not violently dislocated and descends a steep place in a tumultuous manner (which, as already mentioned, is not the case which we consider), I had no hesitation in predicting that the result of the experiment would be confirmatory of my theory, and contradictory of the other; that the transverse line would be found to become a continuous curve, and that no other system of fissures could be found in the glacier satisfying the mechanical postulate of the greater velocity of the central parts of the glacier, than the *ribboned structure* of the ice, which I had already pointed out as resulting from a forced separation of the semi-rigid ice, at a vast, though finite, number of points in the breadth of the glacier, and which I showed to exist exactly in the direction required for releasing the mass from the tension induced by the gravity of its parts.

Having gone to Chamouni a few days later, I looked out for a place where the ice should be as compact as possible, wholly devoid of open fissures, and if possible continuous up to the bank. This latter condition I found it impossible to fulfil on the Mer de Glace, at least without ascending to the *névé*, which might be objected to as less rigid than the glacier proper. The former condition was well-satisfied in a sort of bay on the west side of the Mer de Glace between the Angle and Trelaporte, exactly under the little glacier of Charmoz. The part adjoining the western shore of the glacier is indeed highly crevassed, and therefore unfit for this experiment; but at the distance of fifty or sixty yards from the moraine it becomes remarkably flat and compact for a space of about seventy yards in width, and several hundred yards in length, throughout which space there is not a single open crevasse. Now this compact area of ice presents the veined structure in a nearly longitudinal direction, with a degree of delicacy and distinctness not to be found in any other part of this glacier (as I had already remarked in my *Travels*, p. 159), and it contains no other trace of a system of longitudinal fissures or lines of separation of any kind, which could render mechanically possible the distortion of this flat compact surface of so great an extent. Now I have always observed that the veined structure near the side of a glacier is best developed where the ice is least crevassed, or the continuity of the mass most perfect; a fact stated and referred to its true cause from the first date of my speculations on the origin of the blue veins, in the following words :—" The veined structure invariably tends to disappear when a glacier becomes so crevassed as to lose horizontal cohesion, as when it is divided into pyramidal masses. Now this immediately follows from our theory;

for as soon as lateral cohesion is destroyed, any determinate inequality of motion ceases, each mass moves singly, and the structure disappears very gradually*." Now the ice at the point in question is the compacted ice which has just passed round the great promontory of Trelaporte, having been rent by numberless chasms, and which is consolidated by pressure in the bay in question, whilst the centre of the glacier being still on the steep is deeply crevassed. The structure of the even ice is continuously striped with a regularity comparable to that of the finest chalcedony for a distance of some hundred feet. This structure must *have been produced on the spot,* since no such perfect structure exists higher up, and if it did, it must have retained all the marks of dislocation due to the formation and reconsolidation of the fissures, which are so numerous and wide as to render the passage of the glacier quite impracticable if we follow the same strip of ice up towards the promontory of Trelaporte. Let it then be recollected that the structure is *produced here,* under our eyes, on the very spot where the experiments about to be detailed were made, and that the structure in question produced a vertical slaty cleavage so distinct, that the ice broken into hand specimens may be split parallel to it like any slaty rock, and that the fine hard laminæ projecting vertically after the glacier has been washed by rain, permitted the blade of a knife to be thrust between them to a depth of several inches, although they are rarely more than a quarter of an inch thick.

I shall now describe the actual measurements made upon the glacier in order that my method of proceeding in similar cases (when I have only published results) may be understood.

The general position of the experimental surface will be understood from the topographical sketch (Plate VIII. fig. 1.) The theodolite was planted at a fixed point on the ice Q, just within the crevassed portion, which intervened between it and the western shore of the glacier. This point of fundamental and constant reference was fixed by an exactly vertical hole pierced with an iron *jumper,* or blasting iron, one inch in diameter, and was frequently deepened in order to preserve the centre as exactly as possible in the same vertical line in the ice. The theodolite was centred over it at every observation by means of a plummet, which nearly filled the cylindrical hole and permitted an adjustment, which one day with another might be accurate to about one-tenth of an inch. No stick was placed in the hole, but when not in use it was covered by a large flat stone, which effectually prevents congelation in ordinary weather†. The adjustment of the theodolite on the ice is always a matter for patience, but I succeeded in rendering it perfectly stable when once erected by inserting the three feet in cavities in the ice, and filling them carefully with ice chips.

* Third Letter on Glaciers, Edinburgh Philosophical Journal, October 1842, and Appendix to Travels, 1st edit. p. 407.

† On one occasion this precaution having been neglected (in the case of a different mark on the ice), the hole was found completely frozen up after exposure to a day or two of severe weather in the month of August. It was however recovered by observing the beautiful stellar form of the ice-crystallization.

The theodolite, placed at Q, was pointed with its vertical wire on the well-defined angle of an erratic block Q1 on the opposite eastern bank of the glacier, above Les Echellets (see Plate VIII. fig. 2.). By causing the telescope to traverse in a vertical circle, a transverse line joining the points Q, Q1 was determined, and several stations were fixed in the compact ice eastwards of Q, at distances from it of 30, 60, and 90 English feet and subsequently at 120 and 180 feet. These were numbered in succession (1), (2), (3), (4), (5), and the permanence of their positions in the ice was secured as before by carefully driving vertical holes two feet deep, which were occasionally deepened, and covered with flat stones when not in use. As these points were in succession nearer to the centre of the glacier, they were expected to move with gradually increasing velocity in advance of the imaginary line Q, Q1 drawn across the ice.

But as the theodolite stationed on the glacier at Q must partake of its motion whilst the mark Q1 on the bank remained at rest, the visual line QQ1 would appear to revolve *towards* the origin of the glacier, and hence the relative advance of the points (1), (2), &c. would seem too rapid. To estimate the correction for this error the velocity of the glacier at Q must be determined, and also the distance QQ1. For the former purpose the following method was adopted. When an observation at station Q had been completed, by pointing the telescope on Q1 and observing the apparent advance of the points (1), (2), &c., the telescope was reversed in the Y's, or turned 180° towards the western moraine, upon which it indicated from day to day a new position, owing to the angular revolution of the line joining the fixed point Q1 and the moveable point Q. The point Q2 in the topographical sketch (Plate VIII. fig. 1) indicates the point where the visual line touched the moraine at the commencement of the observations on the 9th of August 1844. By the application of a scale or a similar method, the apparent advance of Q referred to the moraine Q2 was regularly measured. It is thus obvious that these apparent motions were *too great* (by the property of diverging lines) in the ratio of the distance Q1.....Q2, to Q1.....Q : and hence it became necessary to ascertain the position of Q2 as well as Q1. For this purpose a base-line of 300 feet was measured on the ice parallel to the length of the glacier, or perpendicular to the transverse visual line, extending from the point marked (3) to the point Q3 in fig. 3, whence by the theodolite the following bearings were taken :—

$$
\begin{array}{lcc}
& \circ & ' \\
Q1 . \; . \; . \; . \; . \; . & 0 & 0 \\
(3) . \; . \; . \; . \; . \; . & 83 & 6 \\
Q2 . \; . \; . \; . \; . & 148 & 0 \\
\end{array}
$$

From which we deduce the distance from

$$
\begin{array}{lll}
Q1 \text{ to } (3) & . \; . \; . & 2479 \text{ feet.} \\
(3) \text{ to } Q2 & . \; . \; . & 640 \text{ feet.} \\
\end{array}
$$

But as (3) is 90 feet east of Q, we have

$$Q.....Q1 = 2569 \text{ feet.} \qquad Q.....Q2 = 550 \text{ feet.}$$

The apparent motion of Q measured on the moraine is greater than the true motion in the ratio $\frac{2569+550}{2569}$ or $\frac{6}{5}$ nearly. The actual motion of Q is readily deduced as well as the apparent rotation of the visual line QQ1. Thus during 16·75 days, the duration of the experiment, the apparent advance of Q referred to the moraine Q2 was twenty-three feet six inches, which at the distance Q1.....Q2 (3119 feet) subtends an angle of 25′ 54″, or almost exactly *one and a half minutes daily.*

The motion of Q during the interval of any two observations of the marks (1), (2), (3), &c. being thus known, the correction applicable to the apparent advancement of the said marks beyond the visual line is at once found by the proportion

$$Q.....Q1 : Q\text{'s motion} :: Q.....(1) : E,$$

where E is the error of apparent position of mark (1). Thus, suppose the apparent motion on the moraine Q2 to be seventeen inches; this, reduced in the ratio of 6:5, gives 14·2 inches for the progress of Q. If the effect on the apparent place of a mark ninety feet from Q were required, we should have 2569 : 14·2 :: 90 : 0·50 inch.

I shall first detail the observations on the total motion of the glacier at Q, during the period to which the experiment extends, with the corrected daily motion and a memorandum of the state of the intervening weather, which accounts by its excessive variability for the remarkable variation of the progress of the glacier*.

TABLE I.

Date.	Interval.	Apparent motion from commencement.	Corrected daily rate.	Weather.
	days.	ft. inch.	inch.	
August 9. 6 P.M.	0	0 0		9. Fine.
10. 3½ P.M.	0·90	1 5	15·6	10. Some rain. 11. Some rain.
12. 2 P.M.	1·94	4 0	13·2	12. Some rain. 13. Snow storm. 14. Some rain. 15. Snow storm. 16. Wet.
17. 2½ P.M.	5·02	11 6½	14·8 {	17. Fine; melting snow on glacier. 18. Showery. 19. Fine; snow still on glacier. 20. Very fine. 21. Very fine.
20. 1 P.M.	2·94	16 0	15·0	22. Rain.
23. 2 P.M.	3·04	19 10	12·5	23. Showery. 24. Cold rain. 25. Fine.
26. Noon.	2·92	23 6	11·8	26. Fine. Glacier dry.

The first three marks on the ice, those placed thirty, sixty, and ninety feet nearer the centre of the glacier than Q, were fixed on the 9th of August, the mark (4) at 120 feet was planted on the 17th, and the mark (5) at 180 feet, on the 19th. The following are the observations on the *apparent* motions of these points past the transversal line through Q, as well as these relative motions corrected for the real movement of the station Q, as explained in last page. To avoid an illusory appearance of accuracy, the results are given to the nearest twentieths of an inch, which is below the possible errors of observation.

* See Travels in the Alps of Savoy, &c., p. 148.

TABLE II.

Table of the *Apparent* and *True* Motions of the Stations *relatively to Station* Q.

Date.	Interval.	Motion of Q.	Apparent relative motions.					True relative motions.				
			30 feet.	60 feet.	90 feet.	120 feet.	180 feet.	30 feet.	60 feet.	90 feet.	120 feet.	180 feet.
	days.	inch.	inch.	inch.	inch.	inch.	inch.	inch.	inch.	inch.	inch.	inch.
August 9. 6 P.M.	0·0	0·0	0·0	0·0	0·0			0·0	0·0	0·0		
10. 3½ P.M.	0·90	14·0	0·8	1·9	2·3			0·65	1·6	1·8		
12. 2 P.M.	1·94	25·6	1·7	3·1	3·7			1·4	2·5	2·8		
14. 1 P.M.	1·96	29·0	1·5	2·8	4·3	0·0		1·15	2·1	3·25	0·0	
17. Noon............	2·96	44	2·1	4·0	6·0	7·1	0·0	1·6	2·95	4·45	5·0	0·0
19. 5 P.M.	2·21	33·2	1·7	2·7	4·4	6·2	8·5	1·4	2·1	3·25	4·75	6·15
23. 1 P.M.	3·83	48	2·5	5·8	7·7	10·3	15·0	1·95	4·7	6·05	8·15	11·75
26. Noon............	2·96	34·6	2·1	4·6	6·5	8·0	11·0	1·7	3·6	5·3	6·4	8·6
Motion from 9th to 26th...	16·76	228·4*						9·85	19·55	26·9		
Motion from 17th to 26th...	11·96	159·8						6·65	13·35	19·05	24·3	
Motion from 19th to 26th...	9·00	115·8						5·05	10·4	14·6	19·3	26·5

The results in the preceding Table have been divided into three periods, corresponding to the unequal times of observation of the last two and the first three stations. The first line of addition includes the motion at thirty, sixty, and ninety feet for nearly seventeen days; the second sum contains the comparative results for four stations throughout twelve days, and the last line contains the entire relative motions of five stations for nine days. These results may be further analysed, as in the following Table, which exhibits the mean daily motion in inches corresponding to each point for these distinct periods, and also the ratio of the relative motion of each point to the actual motion of the glacier at Q, or the zero point, during the same interval.

TABLE III.

Interval in days.	Actual motion of Q, or zero point.	(1.) 30 feet.		(2.) 60 feet.		(3.) 90 feet.		(4.) 120 feet.		(5.) 180 feet.	
		Mean daily relative motion.	Ratio to actual motion.	Daily relative motion.	Ratio to actual motion.	Daily relative motion.	Ratio to actual motion.	Daily relative motion.	Ratio to actual motion.	Daily relative motion.	Ratio to actual motion.
	inch.	inch.		inch.		inch.		inch.		inch.	
16·76	228·4	0·59	0·043	1·17	0·086	1·60	0·118				
11·96	159·8	0·56	0·042	1·12	0·084	1·59	0·119	2·03	0·152		
9·00	115·8	0·56	0·044	1·16	0·090	1·62	0·126	2·16	0·167	2·94	0·229

This Table shows, first, in a striking point of view, the regularity of action of the law by which the variable motion of the different transversal points in the glacier is governed, since the movement in the different intervals bears so near a proportion, that when estimated in terms of the actual motion of the glacier at the place, the relative motion of the parts scarcely differs by unity in the second place of decimals, and is generally much under it. Taking into account the inevitable errors of observation and the extraordinarily unfavourable circumstances of the weather, it is in the very highest degree improbable that this *law of continuity* of the *partial* motions can be ac-

---

* The true sum ought to be about four inches greater. The difference arises from the impossibility of estimating the correct velocities for the fractional intervals.

counted for by any casual justling or sliding of one finite portion of the ice past another, which would inevitably have left some of the points relatively at rest during some one of the many intervals of observation, and given to others evidence of a starting motion until friction had established a fresh position of repose amongst the struggling masses.

Secondly. This Table enables us to establish not only the continuity of motion of any one point, but the continuity of the relation which connects the points (1), (2), (3), &c. For instance, the relative motions of (1) being

·59   ·56   ·56,

and those of (2) being

1·17   1·12   1·16,

the ratios are

1·98   2·00   2·07.

In like manner the ratios between (3) and (2) will be found to be

1·37   1·42   1·44.

Thirdly. The flexure of the ice may be conveniently represented by a diagram, in which the several ordinates are set off corresponding to the relative spaces moved over. But to find the initial positions of the fourth and fifth marks, the proportional motion for the first period, when they were not observed, must be deduced from the comparative velocity of the period when the observations were comparable. Thus by Table II. the relative velocity of (3) to (4) during the time that they were observed together is 19·05 : 24·3; consequently whilst (3) moved over 26·9 inches (4) would have moved over 34·3 inches; the proportional motion for 16·75 days. In like manner for the mark (5) we have the simultaneous motions of (3) and (5) expressed by 14·6 inches and 26·5 inches, and hence by proportion, as before, we find

14·6 : 26·5 = 26·9 : 48·8 inches,

the relative motion of (5) in 16·75 days.

From these data the simultaneous relative motions of these six stations may be projected in a curve, or rather polygon, as shown in Plate IX. fig. 1. This is interesting, as showing very plainly, not only the regulated increase of swiftness of the glacier towards the centre, but that the *variation of the variation* is clearly brought out, indicated by a convexity in the direction of the motion, and confirming the general principle long ago announced by me, that the retardation is relatively greatest towards the side and less towards the centre. I appeal to any one conversant with the laws of mechanics in their practical application, whether the manifest continuity of such a law does not plainly include a continuity in the mutual action of the parts of the mass under experiment, and even independent of the manifest absence of great dislocations, would not establish the doctrine of a molecular yielding, or plasticity in the ice as opposed to the irregular justling of great blocks, admitting that such could exist unperceived.

The period through which this experiment extended (seventeen days) is conclusive against the idea that a small flexure could take place until the accumulated strain on the solid produced a rupture, which relieved the strain, and so forth, *per saltum*. The continuity of glacier motion in every case except that of precipitous descents or ice-falls, first proved by my experiments in 1842, is now universally admitted by those who have had any personal experience in the measurement of glacier motion, however opposed to my theoretical views*. The changes for seventeen days were connected (as has been shown) by a law of continuity established by numerous intervening observations; and the flexure or distortion of the ice amounted in this time to no less than *four feet* at the opposite ends of a line 180 feet in length. It is quite certain, from my own previous observations and those since made by M. AGASSIZ's directions on the glacier of the Aar, that the movement thus shown to have continued seventeen days without a *saltus* would have continued the whole season in the same manner. In fact, the *deformation* or flexure thus observed being sufficient to account for the whole excess of the central above the lateral motion, is in itself an explanation, and a proof that the explanation is adequate, and leaves nothing residual to be accounted for by *saltus*†.

I have more to add on this subject, but shall first give an account of an extension of this experiment on the actual flexure of the ice, upon so elaborate a scale as I scarcely ventured to hope would prove successful, especially as the time I could devote to watch its progress was small, and the circumstances of weather excessively unfavourable.

Having succeeded so well with the thirty feet station in the transverse line, I thought of multiplying the points of observation still further, so as to obtain a polygon of flexure more nearly approaching to a curve. This I did by making the first ninety feet of the transverse line, *i. e.* the space between Q and (3), Plate VIII. fig. 3, the subject of more immediate experiment, fixing in it forty-five stations only two feet apart. After several partial failures, which gave me, notwithstanding, encouraging results, I selected this plan. A space a foot wide and ninety feet long was cleared with hatchets and ice tools, so as to arrive at a nearly even surface of the hard delicately veined ice; and gutters were made so as to drain as far as possible the surface water from the part under experiment. The theodolite being stationed as usual over Q, and the vertical wire of the telescope describing a great circle passing through the line QQ1 transverse to the glacier, an assistant (BALMAT), directed by my signals, bored

---

* See proofs cited in my Ninth Letter on Glaciers, Edinburgh Philosophical Journal, April 1845, and in second edition of Travels in the Alps, Appendix.

† Mr. WILLIAMSON, Fellow of Clare Hall, Cambridge, to whom I proposed this experimental test of the theory of movement by *echelons*, made a series of independent observations on the Mer de Glace, which coincided in result with what has been stated above. After a patient examination of these facts, and of others which he observed on different glaciers, I am glad to say that Mr. WILLIAMSON was led to abandon the theory of sliding columns or fragments, and to accept that of plasticity as connected with the mechanism of the veined structure which I have endeavoured to illustrate above.

a series of holes from two feet to two feet, forty-five in number, with a common carpenter's centre-bit, and as nearly as possible in the visual line. The holes, which were $\frac{5}{16}$th of an inch in diameter and about five inches deep, were immediately occupied by wooden pins prepared for the purpose. These pins were placed as nearly as possible in the visual straight line, but from the nature of the operation some errors were inevitable. The amount of these errors of position or zero of the marks was immediately determined by causing the vertical wire again to traverse the series, the assistant placing over the centre of the head of each pin in succession the zero point of a scale of inches divided both ways, and held parallel to the length of the glacier, so that (the divisions to tenths of an inch being very plainly marked, and divisible by estimation by the telescope) the fundamental position of each pin was determined, and considered as $+$ if in advance of the transverse line (in the direction of the glacier's motion), and $-$ if behind it (or nearer the origin of the glacier). The mere error of reading did not in any case exceed $\frac{1}{20}$th of an inch, though the uncertainty of centring of the theodolite over **Q** might amount to $\frac{1}{10}$th of an inch, or even more. The two marks nearest **Q** had their positions determined by a thread stretched from the station-pointer of the theodolite to the third mark, their distance being too small to be distinctly seen by the telescope.

The very same process, as regards the placing the zero of the scale on the head of the pin and reading off, was repeated on subsequent days, and the new readings *minus* the fundamental readings gave the *apparent* relative motion in the interval. This apparent motion had to be corrected, exactly as before explained, for the rotation of the visual line due to the translation of the fundamental point **Q**.

The following Table contains—(1.) the original readings on the four days of experiment, namely—

1844. August 20. 10 A.M.

August 21. 6 P.M.

August 23. 1 P.M.

August 26. 11 A.M.

(2.) The differences from the fundamental readings or total apparent displacements for each day, reckoning from the commencement. (3.) The same corrected for the rotation of the visual line from the following data :—

August.	Interval. days.	Motion of Q. inches.	Correction at dist. 90 feet. inches.
20 to 21.	1·33	16	0·56
20 to 23.	3·12	38	1·33
20 to 26.	6·08	73	2·56

TABLE IV.—Showing the Apparent and True relative motions of forty-five points two feet apart, in a line transverse to the axis of the Mer de Glace, 1844.

Mark, No.	Readings of Position.				Apparent Motion.			Corrected Motion.			Remarks.
	Aug. 20, 10 A.M.	Aug. 21, 6 P.M.	Aug. 23, 1 P.M.	Aug. 26, 11 A.M.	August 20, 21.	August 20—23.	August 20—26.	August 20, 21.	August 20—23.	August 20—26.	
1.	+0·15	−0·1	−0·25	0·0	−0·25	−0·4	− 0·15	−0·25	−0·45	−0·15	Between 1 and 2 a slight fissure.
2.	+0·3	−0·5	−0·25?	− 0·2	−0·8	−0·55	− 0·5	−0·8	−0·6	−0·55	
3.	+0·15	+0·15	+0·6	+ 0·7	0·0	+0·45	+ 0·55	−0·05	+0·35	+0·4	
4.	−0·2	−0·1	+0·25	+ 0·3	+0·1	+0·45	+ 0·5	+0·05	+0·35	+0·3	
5.	−0·25	−0·35	−0·05	+ 0·3	− 0·1	+0·2	+ 0·55	−0·15	+0·05	+0·25	
6.	−0·25	0·0	+0·6	+ 1·2	+0·25	+0·85	+ 1·45	+0·2	+0·65	+1·1	
7.	−0·2	+0·2	+0·9	+ 1·8	+0·4	+1·1	+ 2·0	+0·3	+0·9	+1·6	Veined structure very strong.
8.	0·0	+0·3	+1·2	+ 2·45	+0·3	+1·2	+ 2·45	+0·2	+0·95	+2·0	
9.	0·0	+0·5	+1·45	+ 2·5	+0·5	+1·45	+ 2·5	+0·4	+1·2	+2·0	Between 9 and 10 a fissure.
10.	−0·2	+0·15	+1·1	+ 2·3	+0·35	+1·3	+ 2·5	+0·25	+1·0	+1·95	
11.	−0·4	+0·25	+1·35	+ 2·7	+0·65	+1·75	+ 3·1	+0·5	+1·4	+2·45	
12.	−0·0	+0·5	+1·6	+ 2·85	+0·5	+1·6	+ 2·85	+0·35	+1·25	+2·15	
13.	−0·55	+0·15	+1·45	+ 3·3	+0·7	+2·0	+ 3·85	+0·55	+1·6	+3·1	
14.	−0·55	+0·35	+1·5	+ 3·7	+0·9	+2·05	+ 4·25	+0·75	+1·65	+3·45	
15.	−0·15	+0·6	+2·05	+ 4·15	+0·75	+2·2	+ 4·3	+0·55	+1·75	+3·45	
16.	−0·2	+0·65	+2·2	+ 4·5	+0·85	+2·4	+ 4·7	+0·65	+1·9	+3·8	
17.	−0·25	+0·6	+2·2	+ 4·15	+0·85	+2·45	+ 4·4	+0·65	+1·95	+3·45	
18.	−0·15	+0·7	+2·4	+ 4·85	+0·85	+2·55	+ 5·0	+0·65	+2·0	+4·0	
19.	−0·1	+0·95	+2·5	+ 5·35	+1·05	+2·6	+ 5·45	+0·8	+2·05	+4·4	
20.	−0·15	+1·0	+3·0	+ 5·9	+1·15	+3·15	+ 6·05	+0·9	+2·55	+4·9	Two fissures between 19–20 and 20–21.
21.	+0·25	+1·7	+3·55	+ 6·7	+1·45	+3·3	+ 6·45	+1·2	+2·7	+5·25	
22.	+0·1	+1·6	+3·5	+ 6·7	+1·5	+3·4	+ 6·6	+1·25	+2·75	+5·35	
23.	+0·3	+1·7	+4·0	+ 7·55	+1·4	+3·7	+ 7·25	+1·1	+3·0	+5·95	22–23, slight fissure very oblique.
24.	+0·2	+1·65	+4·0	+ 7·9	+1·45	+3·8	+ 7·7	+1·15	+3·1	+6·35	
25.	−0·3	+1·15	+3·5	+ 8·0	+1·45	+3·8	+ 8·3	+1·15	+3·05	+6·9	
26.	+0·35	+1·7	+4·3	+ 7·95	+1·35	+3·95	+ 7·6	+1·05	+3·2	+6·1	26–27, a slight fissure. It recrosses the line at 40–41.
27.	+0·25	+1·75	+4·4	+ 8·45	+1·5	+4·15	+ 8·2	+1·2	+3·35	+6·65	
28.	+0·25	+2·05	+4·9	+ 9·1	+1·8	+4·65	+ 8·85	+1·45	+3·8	+7·25	
29.	+0·4	+2·2	+5·15	+ 9·35	+1·8	+4·75	+ 8·95	+1·45	+3·9	+7·3	
30.	+0·15	+1·9	+4·9	+ 9·3	+1·75	+4·75	+ 9·15	+1·4	+3·85	+7·45	
31.	0·0	+2·0	+4·7	+ 9·0	+2·0	+4·7	+ 9·0	+1·65	+3·8	+7·25	
32.	+0·2	+2·3	+5·3	+ 9·7	+2·1	+5·1	+ 9·5	+1·7	+4·15	+7·7	
33.	+0·2	+2·1	+5·5	+10·0	+1·9	+5·3	+ 9·8	+1·5	+4·3	+7·95	
34.	+0·9	+2·9	+6·0	+10·95	+2·0	+5·1	+10·05	+1·6	+4·1	+8·15	
35.	+0·25	+2·2	+5·7	+10·5	+1·95	+5·45	+10·25	+1·55	+4·4	+8·25	34–35, a very slight fissure.
36.	+0·1	+2·1	+5·5	+10·15	+2·0	+5·4	+10·05	+1·55	+4·35	+8·0	
37.	−0·1	+2·05	+5·3	+10·5	+2·15	+5·4	+10·6	+1·7	+4·3	+8·5	
38.	0·0	+2·1	+5·9	+11·3	+2·1	+5·9	+11·3	+1·65	+4·75	+9·15	
39.	+0·2	+2·4	+6·35	+11·8	+2·2	+6·15	+11·6	+1·75	+5·0	+9·4	
40.	0·0	+2·3	+5·9	+11·1	+2·3	+5·9	+11·1	+1·8	+4·7	+8·85	
41.	+0·1	+2·5	+6·4	+12·15	+2·4	+6·3	+12·05	+1·9	+5·1	+9·7	40–41. See 26–27.
42.	0·0	+2·3	+6·25	+11·8	+2·3	+6·25	+11·8	+1·8	+5·0	+9·4	
43.	−0·1	+2·2	+6·15	+11·65	+2·3	+6·25	+11·75	+1·8	+4·95	+9·3	
44.	0·0	+2·4	+6·5	+12·35	+2·4	+6·5	+12·35	+1·85	+5·2	+9·85	
45.	−0·2	+2·4	+6·1	+12·25	+2·6	+6·3	+12·45	+2·05	+4·95	+9·9	

The results of these observations will be best understood by consulting the projected results in Plate IX. fig. 2, in which the advance of the marks is magnified twofold relatively to the spaces between them, which was necessary for the reduced scale of the engraving, although it has the disadvantage of increasing the deviations from symmetry which might arise from errors of observation. As those who have not attempted the actual execution of such experiments cannot be aware of the difficulties they entail, it may be just to mention that during weather so unfavourable as that which occurred during the continuance of these experiments, nothing can be so irksome as the necessity of persevering in the face of physical obstacles, the only alternative which the necessary limitation of my stay afforded being to abandon them. I do not speak of the painful effort of conducting delicate observations for hours under a hot sun, whilst the feet are immersed in the liquid sludge of decaying snow, for I am not aware of having sacrificed the precision of a single observation to such a cause (though in the course of my glacier experience I have sometimes been compelled to abandon or discontinue observations), but it is easy to see that the success of experiments like this depended upon the absolute fixity of the marks inserted in the unstable and wasting surface of the glacier, and that the most dry and uniform condition of the ice seemed alone to promise a chance of finding the small pins in the exact positions in which they had been planted a day or two before. Instead of this, eleven out of nineteen days which I spent at Chamouni were wet, and notwithstanding the season of the year, the glacier was repeatedly covered with snow, which in melting under a succeeding fierce sun, left the surface honeycombed by infiltration and streaming with wet, so that the preservation of the holes was only effected by laboriously covering every one with large flat stones during the intervals of observation, and even this was not free from other disadvantages which it would take too long to particularize. On the whole, the uniformity of the triple curves exhibited in fig. 2 is surprising, considering the local errors to which the fixation of the pins was liable, and the smallness of the quantities sought. The first two curves, those for the 21st and 23rd August, are indeed as perfectly regular as it would be possible to expect from this kind of observation, even much more so than I had ever hoped to attain; but on the 26th the holes containing the pins were more degraded, and some manifest errors have arisen from this cause, and evidently affect only single marks, such as the twelfth and twenty-fifth mark, which singly have inclined forward or backward by the fusion of the ice. With these preliminaries as to the reasons why the irregularities of these curves should be judged with indulgence, I will state briefly the most apparent general results.

First. The flexure of the ice is proportional, almost exactly to the time elapsed in the intervals of the observations; and it is also graduated from point to point, not *staccato*, as would inevitably have been the case had the relative motion been due to the sliding of finite portions past one another, as in Plate VII. fig. 3. We perceive nothing of the kind.

Secondly. A singular peculiarity strikes the eye, which at first puzzled me, but when the cause was explained, confirms in no slight degree the accuracy of the methods employed and their fitness to reveal the minutest motion of the glacier. The curves of Plate IX. fig. 2, cut the axis, not all exactly at the same point; but on an average this point may be fixed with tolerable accuracy between the third and fourth mark, or seven feet from the fundamental station Q. *The first and second marks moved evidently slower than the point Q, or the zero,* and the progress of the third and fourth is dubious or irregular. The cause of this peculiarity was clearly ascertained *on the spot* to be the existence of two crevasses belonging to the lateral system of crevasses, between which at their thinning out the station Q had been placed, under the idea that their distance was sufficiently great not to affect the motion. The position of these crevasses is shown in Plate VIII. fig. 3, by which it will be seen that they were fifty feet apart in the direction of the length of the glacier, and that a line joining their extremities passed eight feet nearer the centre of the glacier than the point Q, thus almost coinciding with the point of contrary flexure, which was evidently occasioned by a slightly superior advance of the mass of ice on which Q was placed, thus insulated in some degree between the two fissures. This enables us to transfer the origin of our curve to a point of undoubted solidity, namely, the fourth station, from which point it swells with the regularity which has been described, and which establishes the compactness of the ice experimented on in the most convincing manner.

Thirdly. The curves reckoned from the origin, thus experimentally indicated, show in a beautiful manner the convexity in the direction of the glacier motion before alluded to, which is singularly striking, considering the shortness of the space in which it is developed with nearly mathematical precision, being only about $\frac{1}{25}$th of the breadth of the glacier in this place (see ground plan, Plate VIII. fig. 1.). Even an inspection of the curves (Plate IX. fig. 2) can faintly convey the impression made upon my own mind, when upon the 26th of August I placed the theodolite for the last time over station Q, and caused the vertical wire to pass in front of the line of pins bent into the convex shape by the relative motion of six days' continuance. Thus seen in foreshortened perspective, the eye would in an instant have seized an abrupt motion or discontinuity of the line, but " the appearance of the curve they formed was beautiful ; the whole line of pins was deviated from the usual line QQ1 by an angle equal to 12·45 inches, seen at a distance of ninety feet, or about 40′, and besides this, the pins lay in a beautiful and nearly continuous curve, presenting its convexity towards the valley, and decidedly without any great step or start. This was beautifully seen when I directed the vertical wire of the theodolite upon the forty-fifth pin and caused it to describe a vertical plane*. I observed however a curious fact, plainly indicated by the numerical results ; the curve crossed the axis at the fourth pin, and attains its greatest convexity at the twenty-fifth†."

Fourthly. That no information might be wanting as to the precise condition of the mass of ice under experiment, I made a very minute examination of the state of the transverse line with respect to the occurrence of *flaws* in the ice. The most important of these was one which returned into itself, crossing the line towards the origin of the glacier between the twenty-sixth and twenty-seventh marks, and returning backwards between the fortieth and forty-first without extending further upwards. Such a flaw, even if devoid of cohesion, could only act by allowing the piece of ice contained between the twenty-seventh and fortieth mark to slip bodily forwards,

---

* An approximation to this effect will be obtained by stretching a fine thread over the figure.

† From my notes made at the time.

leaving a vacuity behind. No such slip is recorded in the observations, and no such gap or vacuity was left during the continuance of the observations for seventeen days. Nothing approaching to an open fissure occurred in any part of the space of ice under observation, and the few flaws noted (see the column of remarks in Table IV., and the, indications of them by dotted lines and arrows in Plate IX. fig. 2, showing their directions) were perfectly close, and all more or less of zigzag forms, preventing the possibility of sliding. The *veined structure* was developed in every part of the section, in some parts more admirably than others, as near the seventh mark, and between the fortieth and forty-fifth. These countless blue veins may be considered as so many flaws or partial solutions of continuity, existing or having existed; they are almost perfectly *straight*, and (as will be shown immediately) *exactly* in the direction in which the relative motion of the parts of the ice demonstrated by these experiments takes place. It would require strong evidence to convince us that these veins are not occasioned by, and the mechanism of, the plastic motion of the ice.

The beautiful convexity of the curves in the direction of motion which the eye at once seized, and could more accurately distinguish by the theodolite, as having its largest sagitta at the twenty-fifth mark, namely, almost exactly midway between the fourth mark or convex origin, and the forty-fifth or extreme mark, contains in itself an evidence with which no person of correct habits of thought can fail to be struck as proving a regular plastic action of gravity or other propelling force acting from point to point on the mass of the glacier. I made however a check experiment almost unnecessary, but which I will here detail. Lest it should be alleged that the whole area under experiment was a moving one capable of being *swung* round by the pressure upon the centre of the glacier, so that the displacement of the transverse line was due to rotation of the mass operated on round some distant centre, I took care near the commencement of my experiment to fix a mark in the ice in the same line with Q parallel to the length of the glacier, or perpendicular to the transverse visual line at the commencement of the experiment. This point is marked q in Plate VIII. fig. 3. Now had the block of ice on which the marks Q, q, (1), (2), (3), &c. were fixed, been revolving round some fixed or moveable centre, the right angle (3) Qq would have remained a right angle, and so for (2) Qq and (1) Qq. But these angles constantly became more obtuse, and that in different degrees depending on the convexity of the curve Q, (1), (2), (3): and so of (4) and (5). It is impossible, therefore, to avoid the conclusion, that the solid ice was itself *distorted* to the amount of the excess of progression of the more central above the lateral stations.

The results are contained in the following Table, the five stations being those formerly mentioned at 30, 60, 90, 120, and 180 feet from Q. The first three stations were fixed when the visual line had an azimuth of 89° 55′ from q; the fourth was fixed on the 14th of August, when the visual line had revolved through 3′ (its daily progression towards q in consequence of the translation of the station Q being 1′ 30″ exactly), and the fifth on the 17th, when the visual line was in azimuth 89° 47½′.

TABLE V.—Showing the variable azimuths (observed) of the transverse stations with the longitudinal direction Q*q*.

Date.	(1).	(2).	(3).	(4).	(5).	Angle Q1 Q *q*.
	° ′	° ′	° ′	° ′	° ′	° ′
August 12. 2 P.M.	89 55	89 55	89 55	......	......	89 55
14. 1 P.M.	90 6½	90 5	90 5	89 52		
17. Noon.	90 26	90 24	90 23	90 8	89 47½	
19. 5 P.M.	90 47	90 46	90 44	90 10	89 52	
21. 6 P.M.	90 56	90 56	90 53	90 21	90 3	
23. 1 P.M.	91 9½	91 9½	91 4½	90 34	90 14	
26. Noon.	91 24	91 23	91 18	90 45	90 24½	89 28½

In the last column is the observed azimuth of the visual line joining the fixed point Q1 with the moveable station Q, from the mark *q* in the glacier, on the first and last observation. This angle had therefore diminished by $26'\frac{1}{2}$. Now from the known progression of Q, and the known distance QQ1, we computed at page 166 the rotation of the visual line, and found it to be $25'\ 54''$. This close correspondence is highly satisfactory, as showing that the relative positions of Q and *q* remained unchanged during the continuance of the experiment with reference to the imaginary transverse line by which they were adjusted. It is worthy of notice that the line Q*q* was exactly in the direction of the veined structure.

Edinburgh, July 1845.

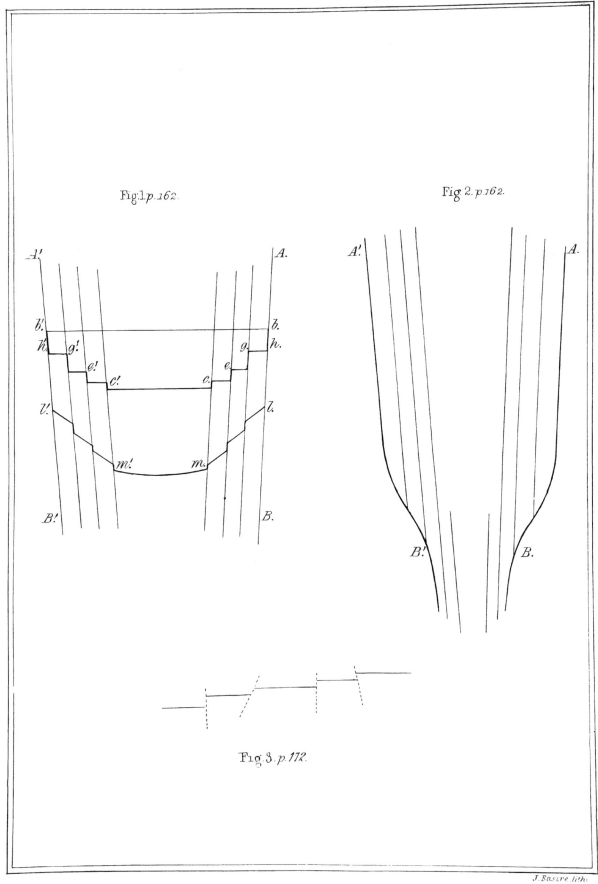

Fig.1.p.162.

Fig 2.p.162.

Fig.3.p.172.

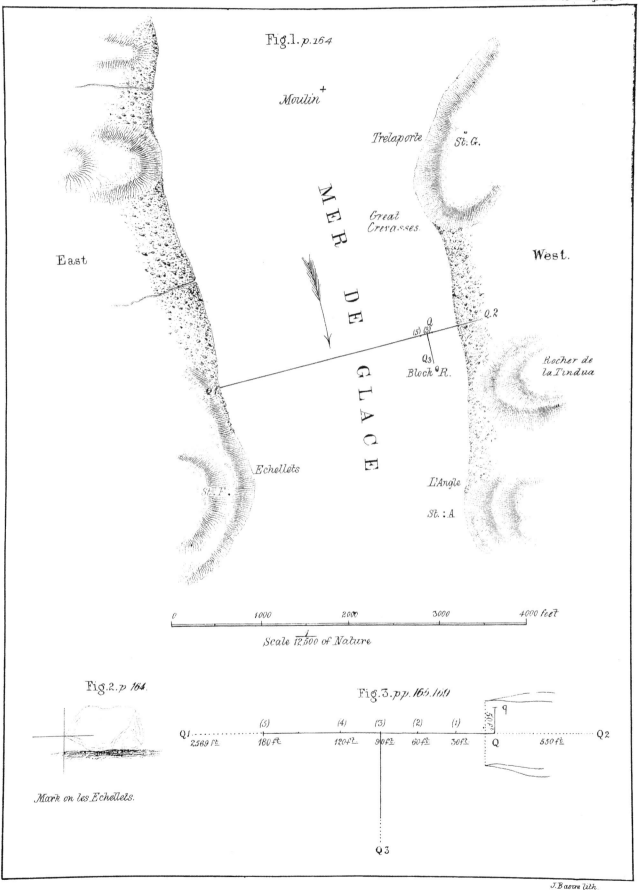

Fig. 1. *p. 164.*

+
Moulin

Trelaporte   *St. G.*

M E R

Great Crevasses.

East

West.

D E

Q.2
(5) (3) *Q*
Q3
Block °R.

Rocher de la Tindua

G L A C E

Echellets

L'Angle

St. F.

St. : A

Q1

0   1000   2000   3000   4000 feet

Scale $\frac{1}{12,500}$ of Nature

Fig. 2. *p 164.*

Fig. 3. *pp. 165.169*

(5)   (4)   (3)   (2)   (1)

q

Q1 .... Q2

2569 ft.   180 ft.   120 ft.   90 ft.   60 ft.   30 ft.   Q   550 ft.

Mark on les Echellets.

Q3

J. Basire lith.

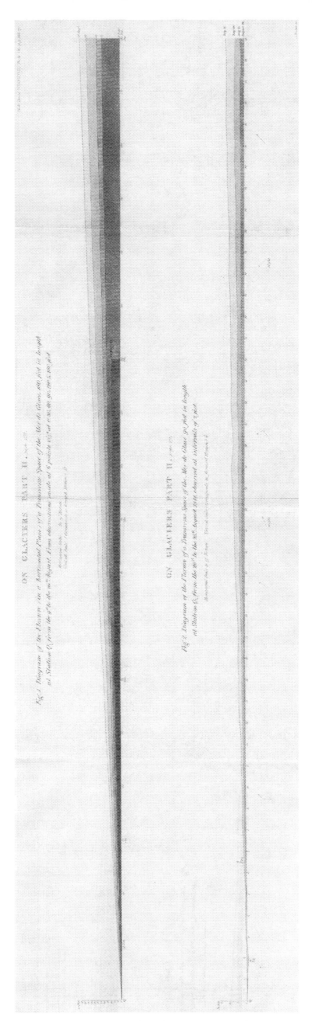

The material originally positioned here is too large for reproduction in this reissue. A PDF can be downloaded from the web address given on page iv of this book, by clicking on 'Resources Available'.

XIV. *Illustrations of the Viscous Theory of Glacier Motion.*—Part III.
By JAMES D. FORBES, *Esq., F.R.S.S. L. and E., Corresponding Member of the Institute of France, and Professor of Natural Philosophy in the University of Edinburgh.*

Received January 27,—Read February 26, 1846.

§ 6. *On the Motion of Glaciers of the Second Order.*
§ 7. *On the Annual Motion of Glaciers, and on the Influence of Seasons.*
§ 8. *Summary of the Evidence adduced in favour of the Theory.*

§ 6. *On the Motion of Glaciers of the Second Order.*

UP to the year 1844 no attempt had been made, so far as I am aware, to measure the rate of motion of those comparatively small isolated glacial masses reposing in the cavities of high mountains, or on *cols*, called by DE SAUSSURE *Glaciers of the Second Order.*

Some observations had indeed been made upon a glacier of this description in 1841, and by MM. MARTINS and BRAVAIS, during a residence on the Faulhorn. But it was not at that time known that the motion of glaciers was a continuous and regular one, admitting of rigorous measurement even in short intervals of time, and the importance of such observations was overlooked. They accordingly believed that the glacier in question had no sensible motion, and probably they did not attempt to observe it until a subsequent year. It is impossible now to doubt that the *Blau Gletscher*, near the Faulhorn, has a movement like all other bodies of the kind.

In July 1844, I had an opportunity of passing some days at the hospice of the Simplon, in the neighbourhood of which exists a small glacier of the second order, easy of access, and very fit for the experiment which I proposed to myself upon such bodies. Its diminutive size made it all the more suitable ; for should it be found to possess a regular motion, we are certain that the *mechanism of a glacier* is continued within the small compass of a mass which may be conveniently examined in detail in all its parts. It is lodged in a niche of the mountain called the Schönhorn*, immediately behind the Simplon hospice : we shall therefore call it the glacier of the Schönhorn. From its inconsiderable extent, it might easily be overlooked by a passing traveller amidst the multitude of vast and striking objects by which he is surrounded†. It is perched, as has been said, in a kind of niche on the northern

* Also called Hübschhorn, an equivalent epithet.

† The reader will not for a moment imagine that it is the Kaltwasser glacier of which we speak, which lies also in the neighbourhood of the Schönhorn, descending from the Monte Leone and Wasenhorn, and from which the *Galerie du Glacier* on the Simplon road takes its name.

face of the Schönhorn, somewhat about an hour's steep climb above the hospice; consequently about 1400 feet higher. The hospice is itself 6580 English feet above the level of the sea; the mean height of the Schönhorn glacier may be taken at 8000 feet. I had not an opportunity of ascertaining it more accurately.

Plate X. figs. 1 and 2, shows a sketch of a front view of the Schönhorn taken from the opposite heights, and a ground plan of the glacier. The latter is sketched merely by the eye, but the scale is furnished by some actual measures. I first visited the glacier on the 20th of July 1844. It was then covered over, by far the greater part of its extent, with snow, as shown in the plan. This snow is in great part manifestly permanent, and the glacier is therefore in the state of *névé*. The general slope is from top to bottom of the plan, and its inclination is variable, depending upon the direction of the avalanches by which it is fed, of which the principal descends the rapid *couloir* marked C, when the inclination is about 35°. This avalanche forms a sort of ridge down the glacier, as indicated by the shading of the map, leaving a considerable space comparatively flat to the eastward. On the west, the snow thins off from the ridge until it exposes the ice near the part marked B, where the slope is still considerable, being 20°, and here we have the real mass of the glacier exposed, although the ice is not of an exceedingly hard or crystalline character. The front or lower termination of the glacier all along presents a steep, nearly precipitous surface of ice, sloping from 45° to 60°. This ice rests on a bed of debris of rock which appears to be inclined about 25°. Except near the precipitous termination of the glacier, there are no apparent crevasses. The surface is uniform and uninterrupted. Some water issues from beneath the steepest part of the ice; but even in the middle of the day, near the end of July, there was exceedingly little. The length, if it may be so termed, of the glacier, from back to front is about 1000 feet, and its greatest breadth 1300 feet. Its surface may be roughly estimated at twenty-six acres.

The rock of which the Schönhorn is composed, is an alternation of the slaty rocks resembling gneiss with talc slate, which are so common in this part of the Alps. To my great surprise, on one of my visits, I heard the sound of hammers and blasting in this elevated and remote spot; and found two men employed in quarrying Pot-stone (*Lapis ollaris*) for building ovens, from a retired nook beyond the glacier; the quarry is marked on the plan at E.

On the 20th of July 1844, I ascended to the glacier, accompanied by M. ALT, one of the clerical members of the Simplon establishment, and an assistant; and I fixed upon a position, marked St. on the rock on the east side of the glacier, for planting the instrument, which was then directed, as nearly as I could judge, in a line transverse to the prevailing slope of the glacier, and the telescope was made to describe a vertical plane. It was then sighted upon a well-marked quartz vein on the rock on the distant side of the glacier, marked D, by which it could at any time be brought into precisely the same position; the position of the instrument itself being referred to a mark cut on the rock where it stood. Two marks were then fixed on the glacier; one was a pole stuck in at A, several feet into the snow of the avalanche already described as traversing the length of the glacier. The slope of the snow at the point A was about 10°; and the distance of A from the station St., by an approximate measurement, 340 feet. 350 feet further in the same direction, a hole was made with a blasting iron into the solid ice at B, where the inclination was 20°. The precise position of these marks being determined relatively to the visual line, the observation was finished at 4 o'clock P.M.

On the 23rd of July we returned. The mark A in the snow (which was so firmly driven in that it could not be withdrawn without breaking the pole) had advanced in the direction of the slope exactly four inches at 1 P.M., or in sixty-nine hours; whilst the mark B in the ice had advanced $5\frac{1}{4}$ inches in the same time; whence we have

Velocity of A in twenty-four hours . . . . 1·4 inch.

Velocity of B in twenty-four hours . . . . 1·8 inch.

The result was what I had anticipated, although it must be confessed it might be expected to be nearly the same upon any theory of glacier motion yet proposed. The slope of a glacier, *per se,* is not an index of what should be the velocity of motion on the viscous theory. No doubt, other things being equal, the velocity will be proportional to some function of the declivity, and such we have seen to be fully borne out by experiments on the Mer de Glace of Chamouni; and in the present case, the velocity under a slope of 20° was about one-third greater than that under a slope of 10°. But the analogy of a river, as well as theoretical considerations, show that the slope is but one of numerous considerations; such as (1) the *mass* of the viscous body; the smaller the mass the smaller the velocity on a given slope*; (2) the state of infiltration or wetness of the glacier altering its resistance to change of form†. Without mentioning other causes, these are quite sufficient to account for the small velocity observed, when we recollect the very insignificant mass of this glacier and its dry state arising from its great elevation, its northern exposure, and even the very inclination of its bed which keeps it in a state of perfect drainage and leaves it always in a state tending to the *snowy,* rather than that of imbibition.

§ 7. *On the Annual Motion of Glaciers, and on the Influence of Seasons.*

The first estimate of the least authority on the advance of any point of a glacier from year to year, was made by Hugi on the glacier of the Aar, from 1827 to 1836. The method employed was to measure the distance of a well-marked block of stone, resting on the ice from a transverse line determined by the fixed objects on the shore. This is the only way, generally speaking, practicable upon glaciers at a distance from habitations, and where marks cannot be conveniently renewed in the ice from time to time during the whole year. The velocity of the part of the glacier imme-

<hr/>

* Travels in the Alps of Savoy, 2nd edit., p. 387. † Ibid. pp. 148. 371.

diately below the promontory, called the Abschwung, was found to be about 240 feet per annum, which though neither confirmed nor invalidated by the discordant measurements subsequently made by other observers on the same glacier, has at length been substantially corroborated by a professional surveyor, M. WILD, who has recently undertaken the verification at M. AGASSIZ's request.

After having myself observed the motion of several points of the Mer de Glace of Chamouni during the summer of 1842, I fixed the positions of two conspicuous blocks, one near Montanvert, marked D 7 ; and another opposite the Tacul, marked C, or the Pierre Platte (see my Map of the Mer de Glace), by means of which I hoped to ascertain the mean annual motion in succeeding years. With respect to the latter, or the Pierre Platte, I was successful; for in September 1843 I ascertained geometrically its change of position, subject however to the uncertainty of a few yards, owing to the sliding of the block from the pedestal of ice upon which it was so picturesquely poised*, a circumstance which happens once or twice in the course of every summer.

From the 17th of September 1842, to the 12th of September 1843,

the advance was (in 360 days) . . . . 256·8 feet.

Or reduced to the exact year of 365 days . . . . 260·4 feet.

Mean daily motion . . . . 8·56 inches.

Again, being enabled to repeat the measures in 1844, I found the advance—

From the 12th of September 1843, to the 19th of August 1844 (342 days) 270   feet.

Proportional motion for 365 days . . . . 288·3 feet.

Mean daily motion . . . . 9·47 inches.

In the case of the block D 7, I was less fortunate. It was very near the western side of the glacier, and though not thrown up on the shore, yet the ice on which it rested got in some manner so embayed or entangled, that though its motion had been steadily watched during the winter of 1842–43 by my able assistant, AUGUSTE BALMAT, it had scarcely moved since his last observation on the 8th of June 1843, when I visited it in September of the same year. It must be presumed that it had been much retarded previously, and hence it is clearly inadmissible to infer a proportional motion for the portion of the year when it had not been observed, as I did in the Postscript at the end of the first edition of my Travels, whilst in ignorance of the then unsuspected retardation. The motion actually observed was 432 feet in 322 days, being at the rate of 483 feet per annum, or 15·88 inches per day. This is therefore undoubtedly *below* the true measure of the annual motion of the side-part of the glacier somewhat in advance of the Châlet of Montanvert (see the position of D 7 in the Map). It may at least be of some service as an *inferior limit* of the annual motion there.

In 1843 I fixed approximately the position of a block marked P, higher up the glacier than the Montanvert, and near its left bank, exactly opposite the spot called Les Ponts.

* See Frontispiece to Travels through the Alps of Savoy.

The observation, being repeated the ensuing year, gave a motion of *about* 486 feet (the nature of the observation did not admit of the same accuracy as at station C) from the 13th of September 1843 to the 9th of August 1844, or 331 days, being at the rate of

536 feet per annum,

or   17·62 inches per day.

In 1844 I made the casual discovery of one of my staves, used to mark the position of the station A at the *Angle*, a little higher up the glacier than the last, a point of which the motion had been most carefully observed during the summer of 1842 (see Travels, p. 140).   This stick still bore legibly written upon it the date when it had been fixed in the ice at station A, and as the painted marks on the rock of the *Angle* were still as fresh as when they were made, I had no difficulty in finding the exact position on the glacier which this mark had in any part of the summer of 1842, and by measuring the distance to the place where it was found (which was on a spot of the ice quite unfrequented by guides or any one else), I had good reason for believing that this must be the space over which it had travelled in the mean time ; although of course I do not ascribe to this observation the weight of a direct measure, yet it proves an interesting confirmation.   Reckoning from the position it occupied on the 1st of September 1842, it had advanced down to the 26th of August 1844, or in 720 days   952   feet,

or, per annum   . . . .   482·5 feet.

Mean daily motion . . .   15·87 inches.

It will be seen that this result is in close agreement with that observed at station P above mentioned, which is a little further down the glacier, but about the same distance from the side ; for though the motion of P is somewhat greater for 1843–44 than the mean motion of A for 1842–44, it will be seen by the comparative observations at C already referred to, that the glacier moved more rapidly in 1843–44 than in 1842–43.

But I am now enabled to present a view of the actual progress of two glaciers during every part of the year from direct observation.   For these I am indebted to the intelligent and persevering zeal of my excellent guide and assistant at Chamouni, Auguste Balmat, of whose character I have had the pleasure of forming a more and more favourable estimate the longer I have been acquainted with him.   To the long training of the laborious summer of 1842, when he assisted me, he adds the further experience derived from my visits in 1843 and 1844, in the latter of which especially he became familiar with the nice precautions requisite in conducting the most accurate measurements, and received instructions from me which rendered him perfectly competent to continue by himself the simpler kind of measurements which I have alone required of him.   The extraordinary exertions which he used to obtain the winter motion of the block D 7, under the Montanvert, in 1842–43, have been noticed in my former publications.   On one or two occasions, as I learned afterwards from himself, being unable to ascend the usual path to the Montanvert for fear of spring avalanches, he actually clambered with a companion up the rugged ascent from the source of the Arveiron, plunging continually up to the middle in snow, for no other purpose than to make the observation which I had requested of him; and it would be unjust not to mention at the same time the admirable, because rare, generosity, with which he positively refused for himself any share of the remuneration which I pressed upon him the following summer, as some recompense for the fatigues and dangers which he had braved to obtain for me this information. With such a person, my confidence in the observations which he has since made at points much more accessible, and with the experience of some additional years, is complete. I do not mean that mistakes may not occur, or even that the measures may not be less exact than I might have taken myself; but from my knowledge of the man, I am nearly as confident in their *being faithfully reported, exactly as they were made,* as if I had done so myself.

With a view to lighten the labour as much as possible, I selected two stations on the *glacier of Bossons,* and desired BALMAT to select two on the *Glacier des Bois* (the outlet of the *Mer de Glace* towards the valley of Chamouni); all these points being tolerably accessible at every season of the year.

The general method of observation was the following :—vertical holes were driven into the ice with a 4-foot blasting iron, at the points whose motion was to be determined; and these holes were renewed from time to time as the surface of the ice wasted. A staff of wood $5\frac{1}{2}$ feet long, was stuck in each, which projected sufficiently above the snow (which never appears to have exceeded $2\frac{1}{2}$ feet deep on the glacier) to make it visible at all seasons. During winter the staves were frozen into the ice, and the waste being small, the holes did not require renewal. Two marks are then made of a permanent kind on the rocks of the moraine, or two staves driven in, or a distant object on the farther side of the glacier was observed, so as to mark out sufficiently a line transverse to the glacier, the prolongation of which passes over the hole in the ice when first made; and the advance of the hole in the ice beyond this fixed visual line marks the progress of the glacier. The want of a theodolite is supplied by directing the eye past a plumb-line suspended over the fixed mark on the moraine nearest to the glacier, the eye of the observer being over the farthest mark. As the spaces moved over were in most cases considerable, an error of a few inches, or even a foot, is not important to the result. The progress was in every case determined by means of a line marked with *English* feet and inches, left by me at Chamouni on purpose.

The results were communicated to me regularly by letter at intervals of a few weeks during the whole year, and all questions asked and explanations required by me were answered by return of post.

Those who may look with suspicion upon observations made in a remote place by a peasant of the better class, though they may not partake of my security in the results from knowing the character of the individual, will, I believe, have their doubts removed by the internal evidence of this important series of observations, which even a philosopher could not have invented, and which, it will be seen, are confirmed by data of quite another kind over which the observer could have no control, I mean the Meteorological Registers of Geneva and St. Bernard.

TABLE I.

First Station on the Glacier des Bois, a little way below the Chapeau, and at about *one-third* of the breadth of the glacier from its eastern bank.

From 1844			To 1844.			Space moved over in English feet and inches.		Daily motion.
						feet.	inches.	inches.
October ..	2.	10 A.M.	October....	14.	9 A.M.	32	0	32·0
October ..	14.	9 A.M.	November ..	2.	8 A.M.	43	11	27·8
November..	2.	8 A.M.	November ..	19.	4 P.M.*	34	11	24·2
November..	20.	1 P.M.	December ..	4.	3 P.M.	13	10	11·8
			1845.					
December..	4.	3 P.M.	January	7.	3 P.M.	32	8	11·5
1845.								
January....	7.	3 P.M.	February ..	18.	3 P.M.	49	2	14·0
February ..	18.	3 P.M.	March	18.	2 P.M.	39	10	17·0
March	18.	2 P.M.	April	17.	10 A.M.	42	1	16·9
April......	17.	10 A.M.	May	17.	8 A.M.	56	3	22·5
May	17.	8 A.M.	May	31.	2 P.M.	43	11	37·0
May	31.	2 P.M.	June	19.	4 P.M.	61	11	38·4
June	19.	4 P.M.	July	4.	10 A.M.	52	0	42·3
July	4.	10 A.M.	July	18.	5 P.M.	62	0	52·1
July	18.	5 P.M.	August	6.	4 P.M.	77	6	49·0
August	6.	4 P.M.	October....	8.	4 P.M.	187	8	35·7
October....	6.	9 A.M. (?)	November ..	8.	2 P.M.	100	9	36·4
November..	8.	2 P.M.	November ..	21.	1 P.M.	32	6	30·1

TABLE II.

Second Station, Glacier des Bois, near the lowest extremity, just behind the " Côte du Piget."

From 1844			To 1845.			Space moved over in English feet and inches.		Daily motion.
						feet.	inches.	inches.
December ..	4.	2 P.M.	January	7.	4 P.M.	8	6	3·3
1845.								
January....	7.	4 P.M.	February ..	18.	4 P.M.	8	11	2·6
February ..	18.	4 P.M.	March	18.	3 P.M.	7	0	3·0
March	18.	3 P.M.	April	17.	11 A.M.	11	7	4·6
April......	17.	11 A.M.	May	17.	9 A.M.	18	4	7·3
May	17.	9 A.M.	May	31.	1 P.M.	10	5	8·8
May	31.	1 P.M.	June	19.	2 P.M.	13	3	8·3
June	19.	2 P.M.	July	4.	9 A.M.	14	1	11·1
July	4.	9 A.M.	July	18.	6 P.M.	17	6	14·6
July	18.	6 P.M.	August	6.	3 P.M.	18	8	11·9
August	6.	3 P.M.	October....	6.	7 A.M.	50	1	9·9
October....	6.	7 A.M.	November ..	8.	4 P.M.	27	2	9·8
November..	8.	4 P.M.	November ..	21.	11 A.M.	8	0	7·5

* There is some uncertainty about the circumstances of this observation, which from the difficulty of corresponding satisfactorily at so great a distance about minute local occurrences, I have been unable perfectly to clear up. It is probably correct as it stands.

TABLE III.

First Station on the Glacier des Bossons. Some way above the Plateau, where the glacier is usually crossed; on the west side, and near the moraine.

							Space moved over in English feet and inches.		Daily motion.
From 1844			To 1844.				feet.	inches.	inches.
November..	20.	Noon.	.... December ..	4.	3 P.M.		20	4	17·3
			1845.						
December..	4.	3 P.M.	.... January....	7.	1 P.M.		44	11	15·9
January....	7.	1 P.M.	.... February ..	14.	4 P.M.		46	9	13·6
February ..	17.	4 P.M.	.... March	17.	Noon.		35	10	15·4
March	17.	Noon.	.... April	16.	5 P.M.		32	7	12·9
April......	16.	5 P.M.	.... May	17.	3 P.M.		60	0	23·3
May	17.	3 P.M.	.... May	31.	8 A.M.		49	0	42·9
May	31.	8 A.M.	.... June	19.	10 A.M.		54	2	34·1
June	19.	10 A.M.	.... July	4.	5 P.M.		53	8	42·1
July	4.	5 P.M.	.... July	21.	2 P.M.		43	1	30·6
July	21.	2 P.M.	.... August	7.	3 P.M.		40	10	28·8
August	7.	3 P.M.	.... October....	6.	Noon.		103	0	20·6
October....	6.	Noon.	.... November ..	10.	2 P.M.		56	8	19·4
November..	10.	2 P.M.	.... November ..	22.	11 A.M.		22	5	22·6

TABLE IV.

Second Station on the Glacier des Bossons. Near the lowest extremity of the glacier, where free from the moraine, on the western side.

							Space moved over in English feet and inches.		Daily motion.
From 1844			To 1844.				feet.	inches.	inches.
October....	2.	2 P.M.	.... October....	13.	4 P.M.		12	11	14·0
October....	13.	4 P.M.	.... October....	31.	10 A.M.		25	1	17·0
November..	20.	1 P.M.	.... December ..	4.	2 P.M.		15	4	13·1
			1845.						
December..	4.	2 P.M.	.... January....	7.	3 P.M.		36	11	13·0
1845.									
January....	7.	3 P.M.	.... February ..	19.	1 P.M.		43	1	12·0
February ..	19.	1 P.M.	.... March	17.	1 P.M.		27	10	12·8
March	17.	1 P.M.	.... April	16.	6 P.M.		25	8	10·2
April......	16.	6 P.M.	.... May	17.	4 P.M.		49	11	19·4
May	17.	4 P.M.	.... May	31.	9 A.M.		34	6	30·2
May	31.	9 A.M.	.... June	19.	11 A.M.		44	2	27·8
June	19.	11 A.M.	.... July	4.	6 P.M.		41	2	32·3
July	4.	6 P.M.	.... July	21.	11 A.M.		37	3	26·4
July	21.	11 A.M.*	.... August	7.	4 P.M.		30	3	21·4
August	7.	4 P.M.	.... October....	6.	2 P.M.		82	7	16·5
October....	6.	2 P.M.	.... November ..	10.	4 P.M.		17	3	5·9
November..	10.	4 P.M.	.... November ..	22.	1 P.M.		7	2	7·2

These four sets of observations are projected in Plate XI. fig. 1, where the four lower zigzag curves represent the gradation of diurnal velocity by periods, according to the method adopted in projecting my own observations in my Travels, p. 141. The

* Marked 4 P.M., perhaps by mistake, but computed on that supposition.

general accordance is sufficiently manifest, and the effect of the season of the year is beautifully shown, the following being the minimum and maximum values:—

Daily motion in inches.

Glacier des Bois, No. I., minimum in December . . . . 11·5
Glacier des Bois, No. I., maximum in July . . . . 52·1

Ratio of maximum to minimum . . . . $4\frac{1}{2}$: 1

Glacier des Bois, No. II., minimum in January . . . . . . 2·6
Glacier des Bois, No. II., maximum in July . . . . 14·6

Ratio of maximum to minimum . . . . $5\frac{1}{2}$: 1

Glacier des Bossons, No. I., minimum in March . . . . . 12·9
Glacier des Bossons, No. I., maximum in May . . . . . . 42·9

Ratio of maximum to minimum . . . . $3\frac{1}{2}$: 1

Glacier des Bossons, No. II., minimum in March . . . . . 10·2
Glacier des Bossons, No. II., maximum in June . . . . . . 32·3

Ratio of maximum to minimum . . . . $3\frac{1}{4}$: 1

From these observations we may deduce the annual motion from November 1844 to November 1845 with considerable exactness. Allowing for the fractional parts of a year, we obtain the following results, amongst which I have included a separate computation of the mean daily motion for the summer period (April—October), and the winter period (October—April).

TABLE V.

	Bois, No. I.	Bois, No. II.	Bossons, No. I.	Bossons, No. II.
	feet.	feet.	feet.	feet.
Motion for 365 days, November 1844 to November 1845....	847·5	220·8	657·8	489·1
	inches.	inches.	inches.	inches.
Mean daily motion	27·8	7·3	21·6	16·1
Mean daily motion, summer period, April to October	37·7	9·9	28·0	22·2
Mean daily motion, winter period, October to April	19·1	4·7	15·8	10·7
Ratio, summer : winter, motion...................	2·0 : 1	2·1 : 1	1·8 : 1	2·1 : 1

I. From this Table we deduce in the first place a mean annual motion far greater than has hitherto been observed, or perhaps suspected in any glacier, that of near 300 yards, or almost *one sixth* of a mile. This is on the Glacier des Bois beneath the *Chapeau*, where the inclination of the glacier is very steep, adding a new illustration of the general principle*, that in *similar* circumstances the velocity increases with the slope. To this cause may be added the high temperature of the air of the valley to which in this part of its course it is exposed; but this last cause is alone insufficient; for

II. We find that the lowest part of the same glacier immediately behind the Côte

* Travels, 2nd edit. p. 371.

2 B 2

du Piget, a little way above the source of the Arveiron, and therefore still deéper in the valley, has a mean velocity nearly *four times less*, arising solely from the diminished slope and volume of the glacier in that part*. Hence there must be a condensation of the ice here, a pressure *à tergo*, the quicker moving ice pressing against the slower, consolidating it, remoulding its plastic material and sealing the crevasses; and a slight examination of the state of the glacier at the points in question will show this to be the case.

III. All that has now been said with respect to the two stations on the Glacier des Bois may be repeated with only numerical differences with respect to the two stations on the Glacier des Bossons; the one set of observations confirming the other.

IV. In both glaciers the summer motion exceeds the winter motion in a greater proportion, as the station is lower, that is, exposed to more violent alternations of heat and cold; this we shall find to be general.

Before continuing our deductions, we would call attention to the close relation which may be established between the mean temperature of any portion of the year and the velocity of the glacier corresponding to it. This is done in figure 1, Plate XI., exactly in the same way as I did when comparing my observations in the summer of 1842 with the corresponding changes of temperature†. That is to say, I have projected by *periods* (corresponding to the intervals of observation on the glaciers) the mean temperatures as observed at Geneva and at the Great St. Bernard, which are regularly published in the *Bibliothèque Universelle*, the average of which (separately deduced from the mean of daily maxima and minima, and projected in the upper part of the figure) may represent not inaptly the average temperature to which the glaciers in question, and especially the middle and lower regions of them, are exposed; and further, this average possesses the advantage of being derived from data wholly unconnected with the place or parties where and by whom the observations on the motion of the glaciers were made, and therefore are free from the remotest suspicion of either in any degree influencing the other.

---

* This explains a circumstance which has always hitherto been a difficulty to me; the united testimony of the best-informed inhabitants, not only at Chamouni but elsewhere (as at Zermatt and at the Simplon), to the effect that during winter the lowest end of a glacier, which terminates in a valley, does not greatly protrude, nor force the snow before it. This arises in fact from the comparative smallness of the motion which the *tongue* of such a glacier appears to possess, especially in winter.

† Travels, p. 141.

TABLE VI.

Mean Temperatures (by periods) on the Centigrade Scale, observed at Geneva and the Great St. Bernard*.

	Geneva.		St. Bernard.		Means of Max. and Min.	
	Max.	Min.	Max.	Min.	Geneva.	St. Bernard.
1844. Oct. 2 to Oct. 14. ...	17·37	8·57	4·16	— 1·58	12·97	1·29
Oct. 14 to Nov. 2. ...	11·77	5·13	1·61	— 4·93	8·45	— 1·66
Nov. 2 to Nov. 19. ...	11·84	3·11	1·62	— 5·66	7·47	— 2·02
Nov. 19 to Dec. 4. ...	4·93	— 0·18	— 5·11	—11·25	2·27	— 8·18
Dec. 4 to Jan. 7. ...	1·45	— 2·40	— 5·70	—10·96	— 0·27	— 8·33
1845. Jan. 7 to Feb. 18. ...	1·47	— 4·11	— 6·92	—13·86	— 1·32	—10·39
Feb. 18 to March 18.	4·76	— 2·74	— 3·93	—12·56	1·01	— 8·24
March 18 to April 17.	11·27	1·57	0·19	—10·13	6·42	— 4·97
April 17 to May 17.	16·04	5·56	5·15	— 6·36	10·80	— 0·60
May 17 to May 31...	16·83	6·91	6·04	— 4·67	11·87	0·68
May 31 to June 19...	23·70	12·71	11·48	1·11	18·20	6·29
June 19 to July 4. ...	22·58	12·13	9·81	0·53	17·35	5·17
July 4 to July 18. ...	24·01	12·84	10·20	2·08	18·42	6·14
July 18 to Aug. 6...	23·28	13·20	9·25	1·68	18·24	5·46
Aug. 6 to Oct. 8...	21·15	11·31	7·57	0·10	16·23	3·83
Oct. 8 to Nov. 8...	11·84	3·78	2·38	— 3·70	7·81	— 0·60
Nov. 8 to Nov. 21...	12·54	5·86	— 1·03	— 6·06	9·20	— 3·54

A general comparison of the curves of temperature and those of glacier motion (more particularly on the Glacier des Bois) affords a proof of the justness of the principle laid down by me in 1842, that the motion of the ice " is more rapid in summer than in winter, in hot than in cold weather, and especially more rapid after rain, and less rapid in sudden frosts†;" the evidence of the connection is plainer by mere inspection than any detail could make it. But I request attention to the apparent anomalies of the curves, as affording a stronger evidence of the fidelity with which the measurements have been made, and to the truth of the plastic theory, than perhaps even the general coincidence just referred to.

If the velocity of the glacier depend upon the completeness of its infiltration with water, rendering the whole an imbibed porous mass like a sponge, it cannot depend solely on the mean temperature of any period, but also upon the *wetness* of the surface, whether derived from mild rain, from thawing snow, or from any other meteorological accident which the register of the thermometer cannot of itself indicate‡. Further, a thick coating of snow on the glacier must defend it from the excessive cold of winter just as it defends the earth and plants, and consequently the minimum of motion will not necessarily coincide with the minimum of temperature. Now to

* The last three lines have been added during printing.

† Fourth Letter on Glaciers, Edinburgh Philosophical Journal, Jan. 1843; and Appendix to Travels, 2nd edit. p. 415.

‡ "The *proportion* of velocity does not follow the *proportion* of heat, because any cause, such as the melting of a coating of snow by a sudden thaw, as in the end of September 1842, produces the same effect as a great heat would do."—Travels, 2nd edit., p. 372.

estimate these more irregular causes is not so easy; but some light is thrown upon them by a register of the weather and state of the snow, voluntarily kept for me at Chamouni by AUGUSTE BALMAT; which forms a valuable supplement to the ther-mometrical register of Geneva and St. Bernard. Although the daily details would take up too much space, I will endeavour to give a faithful abstract of them so far as to give a general idea of the climate of Chamouni from October 1844 to November 1845. This diary includes (at my request) occasional notes on the state of the source of the Arveiron, which are of considerable interest.

Weather at Chamouni.

1844. *October.*—A good deal of rain during the month, which on the 10th and 16th fell as snow on the hills (nine inches at Montanvert), and subsequently to the latter day the glacier at the Montanvert was not clear of snow during the winter. 14th. Source of the Arveiron diminished to *one-fourth* (of the summer volume). Ice-vault more than half closed.

November.—Till 14th much rain and snow; fine with frost after. 20th. Source of the Arveiron very low; has not shifted its usual position.

December.—Weather generally fine throughout; cold most severe from 7th to 12th.

1845. *January.*—The weather continued splendid till the 20th; greatest cold from $-2°$ to $-5°$ REAUMUR. 19th. The vault has disappeared at the source of the Arveiron. 20th. The first snow fell which lay at Chamouni, and continued from this day, attaining a depth of $1\frac{1}{2}$ foot in February. Up to this time all the secondary heights, even the Breven and Flegère, were clear of snow, and the weather suitable for Chamois hunting. Occasional snow till the end of the month.

February.—Snow at intervals all the month. 13th. Greatest cold of the season; thermometer $-15°$REAUM. followed by fine weather. 20th. Snow lies $2\frac{1}{2}$ feet deep at the upper stations on the glaciers; $1\frac{1}{2}$ foot at Chamouni. The arch of the source of the Arveiron has wholly disappeared, but the water issues at the usual places as in summer. The water is reduced to a small amount and may easily be stepped across. It is *still whitish and dirty*, though less so than in summer; *except* when a change of weather is threatened (when it is as dirty as in summer)*. *Same date.* The glacier of Bossons has extended itself much. "On ne s'y reconnait presque plus." It is advancing towards the moraine of 1818; and the lower end is at least seventy feet high.

March 1st—3rd, mild, with rain; 3rd—13th, cold; 15th, *heavy rain.* Alternate rain and fine till end of the month. 27th. Not half a foot of snow lying at Cha-mouni. The source of the Arveiron has not opened a vault. The quantity and muddiness of the water the same as at the last report.

---

* This important remark proves that in the middle of winter a temporary rise of temperature of the air over the higher glacier regions (which is the precursor of bad weather) not only produces a thaw there, but finds the usual channels still open for transmitting the accumulated snow water.

April.—First week fine; second week cold with snow; changeable to the end of the month. 16th. Source of Arveiron has not much increased in water since the middle of March. In the end of April the snow first disappeared from the lower part of both glaciers.

May.—The first half of the month fine, with occasional snow; the second half changeable, with rain. 17th. The source of the Arveiron has increased three-fourths (*means probably in the ratio of four to one*) since the middle of April, and is dirty. The ice-vault is not yet formed. 26th. The Glacier des Bossons advances rapidly and is crumbling into pyramids. The end of the glacier is at least eighty-five feet high and advances considerably, particularly during the month of May; and widens greatly.

June.—A changeable and wet month; a very late season*. The snow did not entirely disappear from the Mer de Glace opposite the Montanvert till the beginning of July. 6th—7th. The vault opened at the source of the Arveiron. The quantity of water since the end of May is the usual summer supply.

July.—Commenced with warm weather. 5th. Thermometer 27° REAUM. The snow has disappeared from the ice opposite Montanvert, but some patches remain on the way to the Jardin. The Mer de Glace is much higher in level (about forty feet) than in former years, and the marks made in the rock at the *Angle* (in 1842) are all covered. The crevasses much the same as usual. The glacier of Bossons has also increased greatly, and appears to be approaching its old moraines. The register for the greater part of July has not come to hand.

August.—A very changeable rainy month. 8th or 9th. The arch at the source of the Arveiron fell in, and did not form again during the season.

September.—Also a changeable month. Rain twelve days.

October.—A very fine month. No rain mentioned after the 7th.

A careful examination of this interesting register will explain several of the apparently irregular inflections of the curves of glacier motion. Thus (to continue our general remarks, p. 186) we find

V. At the upper station on the Glacier des Bois the least velocity occurred in December, whilst at the lower station (and at both of those on the Bossons) a minimum coinciding also with that of the temperature of the air took place in January. This coincides with the important fact noted in the preceding register, that the upper part of the Mer de Glace was covered with snow from the 16th of October, which only lay in the valley of Chamouni from the 20th of January; the snow screening the ice from the extremity of the cold.

VI. The comparative march of the two glaciers bears a remarkable relation to their positions and form. In the Bossons we detect at once the sudden transitions and seemingly capricious changes of a torrent; in the Mer de Glace we have the

---

* It will be seen from the temperature curves that the thermometer *fell* considerably in the latter part of June, both at Geneva and St. Bernard.

stately and regulated flow of a river, in which the slighter variations are absorbed
by the predominant inertia of a comparatively stable mass. Now the glacier of
Bossons is, as every one who has seen it knows, a mere icy torrent, " a frozen cata-
ract," which descends in a continuous mass from the level of the Grand Plateau of
Mont Blanc to that of the Valley of Chamouni with very little impediment, with no
confining bulwarks of rock, no contracting straits; and throughout this great vertical
height of at least 9000 feet, the angle of descent is very steep indeed for so vast a
mass. On the other hand, though the part of the Mer de Glace, called the Glacier
des Bois under the Chapeau, is very steep, its "*régime*" is regulated by the supply
derived from the reservoir glacier above, and, precisely as in rivers of great magnitude and length of course, and of moderate declivity, it yields sluggishly to impulsive
or retarding forces which are checked and opposed by the multitude of sinuosities,
the embaying of the ice in rock-bound expansions of the channel, the struggle of its
passage through defiles and the enormous friction of its lower surface. Yet, lest
we might attribute the irregularities of the torrential glacier to causes quite local and
uncertain, we find them reflected more or less distinctly in the movements of the
neighbouring one. Thus the anomalous retardation in the end of March and beginning of April appears in three stations out of four, as does that in the first half of
June, showing clearly that it is not an error of observation. It appears that the
thaw of the winter's snow during the month of May, saturating the pores of the glacier with water, produced (as we know that a thaw always does) a sudden and violent
march, especially of the more susceptible or torrential glacier. So completely had
this sudden move forced on the glacier of Bossons, encumbered by the spring
avalanches and loaded with all the fragments and snow masses which had remained
temporarily suspended during the winter months, that the lower part of the glacier
(as we read in the memoranda to the register) advanced and widened greatly, to an
extent which it had not done for many years past, and seemed to change its whole
character; and in February a similar temporary increase of volume had taken place;
" on ne s'y reconnait presque plus," writes BALMAT; thus accounting for the particular
accession of speed which appears in that month. In both cases, after the rapid march
in February and in May, a reaction takes place; the material is deficient, the excessive pressure has been removed by the previous overflow, and a lull occurs in March
and in June.

VII. These irregularities, such as they are, even should we fail in entirely explaining them, are at least not to be attributed entirely to errors of observation, since different observations (which it is to be recollected were sent to England in so rough a
state that they required to be reduced and computed before the variations of velocity
could be deduced from them) agree amongst one another, and agree with the phenomena casually noted in the Meteorological Register. They are very trifling in the
movement of the Glacier des Bois, which presents a curve of remarkable regularity,
giving a minimum about the end of December, and a maximum in July. The coin

cidence with the curve of temperature is greater throughout than we could have expected, considering the important difference of circumstances which occur in autumn and in spring when the thermometer stands nearly alike, the first chill of autumn depriving the glacier of its fluid pressure more effectually than the severer cold of winter which is tempered by its snowy covering, whilst in spring the first re- laxation of the bands of frost saturates the icy mass with the impetuous streams of melted snow, as effectually as the intensest heat of summer. In fact, the velocity would probably be greatest in spring, were it not that then the ice has attained its greatest consolidation by the slow but continued effect of the winter's cold penetrating its upper layers, though after all probably to no very great depth. But this is undoubt- edly the reason why the minimum and maximum approach so near to one another in point of time in the torrential glacier of Bossons, and it receives an important illus- tration from the independent fact of the observed condition of the source of the Arveiron, which (see the Meteorological Register), though very small in February, was still whitish and dirty before a change of weather, showing that the bands of frost were not so strong as to prevent a temporary relaxation of thaw throughout the mass of the glacier even in winter; and although the *mean* temperature of the air had been rising ever since the middle of January, and the greatest cold had occurred early in February, we find that at the end of March the source of the Arveiron was still as small as in February, and that owing to the coldness of the spring it had not even increased very much till the middle of April, when it almost suddenly resumed its summer volume. Now during all this time the velocities of the glaciers under- went but little change,—some oscillations backwards and forwards,—but took no real start until the frost had given way, and the tumultuous course of the Arveiron showed that its veins were again filled with the circulating medium to which the glacier, like the organic frame, owes its moving energy.

VIII. Being curious to see how far a relation might be established between the temperature of the air and the motion of the glacier independent of the irregularly acting causes above adverted to, I projected in Plate XI. fig. 2, the motions of the several points of the glaciers in terms of the temperature of the air for the periods already mentioned. It is to be recollected, however, that the observations of the thermometer were not made on the spot, and indeed it would have been difficult to have fixed upon a spot which should represent the mean circumstances of the whole glacier. Perhaps, therefore, the average of the observations at Geneva and St. Ber- nard (the mean of whose elevations is 4750 English feet above the sea, and therefore between that of Montanvert and Chamouni) may represent pretty fairly the climateric conditions of the inferior parts of the Glaciers des Bois and Bossons. Now, if we examine the curves of fig. 2, we are struck with *their almost perfect flatness until zero of the centigrade scale of temperature is reached*; but, the thawing point of ice past, the velocity manifestly goes on increasing with the temperature, in a ratio which would appear to be tolerably uniform if we neglect the irregular inflections of the curves.

IX. I am unwilling to multiply deductions which every intelligent reader will draw for himself; but one more I must add. It very clearly appears that the variations of velocity due to season are greatest where the variations of temperature of the air are greatest, as in the lower valleys; but it also appears from Remark VIII., that variations of temperature below 0° centigrade, or 32° FAHRENHEIT, produce almost inappreciable changes in the rate of motion of the ice. Hence, from this circumstance alone, we should deduce that in the higher parts of the glacier (where, for example, it freezes almost every night in summer) the variations of velocity should be least, and indeed comparatively small at different seasons. This is well illustrated by comparing the summer motions of the stations D, A and C, mentioned in the first part of this section, with their annual motion, which exhibit a much slighter excess in favour of the summer period than in the lower stations which we are now discussing. The same thing was observed by M. AGASSIZ's surveyors on the glacier of the Aar, who at first saw, in this not very great inequality, an objection to my theory. On a more searching investigation, however, the objection disappears, as in their later writings they have acknowledged. Their position of observation far up on the glacier of the Aar, in a spot having a mean temperature near the freezing point if not lower, had a summer daily motion of 7·99 inches, and a mean daily motion during the whole year of 6·41 inches*. Now at station C, or the Pierre Platte, on the Mer de Glace, the mean motion for July 1842 was 10 inches, and for the whole year, 1842–43, it was 8·56 inches. It is quite evident that the motion of any point in the midst of a glacier is controlled by that of those which precede and follow it, and that it does not necessarily result, either that all must at once suffer a similar increase or diminution of speed, or that the times of maxima and minima, or even the general form of the annual curve, shall be the same. This leads to an important practical result which we shall follow out in the next section.

§ 8. *Summary of the Evidence adduced in favour of the Plastic or Viscous Theory of Glacier Motion.*

It is often difficult to obtain a calm and full hearing for any new theory or experimental investigation; not because there is any antipathy to novelty, or that experiment is undervalued, but simply because, in an age of bustle and struggle for preeminence, each man is so busy with his own reputation, or the means of increasing it, that he has no leisure to attend to the claims of others; to which may, perhaps, be added, that in the general diffusion of knowledge and acquirement, each reader, finding something in every course of experiment or reasoning which he knew, or thinks he knew before, is apt to run off with the chain of ideas which that one familiar link suggests, and losing patience to follow an argument of which he thinks he can, by his own penetration, anticipate the close. He sits in judgment on errors which are of his own invention, and confronts the author with arguments and opinions already

* Comptes Rendus, Dec. 9, 1844.

thrice refuted and rejected by himself. In an age when all men would be teachers and all write for the press, the lot of an attentive reader falls to few.

I am far from saying that I have been more than usually unfortunate in this respect. But having, like others, seen my opinions disfigured for want of sufficient attention to apprehend them, or the arguments by which they are supported; ignorance of first principles hinted at, and even errors of observation imputed, where it was convenient that such ignorance and such errors should be presumed; I claim the privilege of stating afresh, though very briefly, the leading opinions which I do hold, and some arguments for them, which, if not altogether new, may be placed in a new light.

My chief analogies for the illustration of glacier motion have been drawn from the motion of a river, and by that comparison it in a great measure stands or falls. Slight and partial as is our knowledge of the mechanics of imperfect fluids, the explanation which I have given is founded upon that knowledge, and it appears to me to be sufficiently precise to warrant the inference of an identity of the mechanism in the two cases;—namely, that the movement is due to the internal pressures, arising from the weight of the mass, communicated partly or principally in the manner of hydrostatic pressure throughout a body whose parts are capable of moving or being shoved over one another (by that exertion of force which Dr. Thomas Young calls *Detrusive Force**, which overcomes what is commonly called the Friction of Fluids), so that the velocities vary from point to point of the moving body, being most rapid near the surface and centre, and least so near the banks and bottom.

So viscous fluids move, so bodies (even brittle solids, such as hard-boiled pitch) possessing the ordinary properties of solid bodies often do, if sufficient time and sufficient force be allowed†; the efficiency of time being chiefly this, that a pressure insufficient to produce instant detrusion, will, sooner or later, cause the particles to slide insensibly past one another, and to form *new attachments,* so that the change of figure may be produced without positive rupture, which would reduce the solid to a heap of fragments. This change may either take place without any loss of homogeneity, or by numerous partial and minute rents not everywhere communicating, and therefore not necessarily destructive of cohesion, which may be termed a bruise.

A glacier is not a mass of fragments.—As the analogy of the glacier to a river, in which the fluid principle is greatly in defect, and the cohering or viscous principle is greatly in excess, is the theory which I maintain, it is evident that the analogy of a stream of sand, or loose materials shot from a cart, or any other comparison with an aggregate of incoherent fragments or individual masses, must be wrong if mine be right. And I feel confident, not only that such an incoherent mass could not move after the manner of a glacier, but also that attentive inspection of a glacier at once contradicts such an idea.

* Lectures, I., 135.
† See Professor Gordon's Experiment, Philosophical Magazine, March 1845.

On the *first* point, I maintain that a rugged channel, like that of a glacier, with a moderate slope, being *packed* with angular solid fragments, would speedily be choked, and that farther pressure from behind (for such a mass can only convey thrusts, not strains) would tend to wedge the fragments more tightly. Some grains of dry sand will slide easily down a plate of glass; but try to thrust it forcibly through a narrowing tube, or even a uniform one, the lower end of which rests on a surface over which the sand has poured, and your effort is vain, the tube will sooner burst; and even rocks may be blasted rather than the power of the wedge yield*. If the figure of the bed or channel be in any degree irregular, that is, have expansions and contractions, however smooth its surface, however small the sliding angle of ice upon that surface, the choking of a strait or contraction by the piling of the fragments will be as complete and effectual as if the lateral friction were excessive. Now in point of fact we have such cases as this;—a glacier 2000 yards wide (the Mer de Glace at the Tacul) issues by an orifice or strait 900 yards wide;—the glacier of Talefre, a nearly oval basin, pours out its annual overcharge by an orifice the breadth of which is but one-third of its lesser, one-sixth of its greater diameter†. On the supposition of jostling fragments, the facility of motion is increased, as the comminution is greater. The impossibility of the discharge of a fragmentary solid through a gorge by long stripes fractured parallel to its length, and constituting parallelopipedons of a certain breadth, is evident.

Crevasses.—In the *second* place, I maintain that actual inspection shows that a glacier is not the mass of fragments nor of parallelopipedons which some persons have, naturally enough at first sight, supposed it to be. In truth there is not an approach to such a condition in those glaciers which move over moderate slopes of considerable extent, which have very properly been assumed by all writers as the criterial examples of any theory; for it is not denied that portions of glaciers and glacier tributaries do sometimes fall piecemeal over precipices, each fragment descending by its separate and individual gravity, in the manner of an avalanche, although I am disposed to believe, indeed am sure, that the number of such instances is smaller than is usually imagined; and the angle requisite for such a tumultuous mode of descent is far greater than it has, perhaps, always hitherto been considered to be. To him who would form a just estimate of the mechanical constitution of a glacier—who would consider it as a whole—without *always* distracting his attention from the length and breadth of the problem by a minute attention to its lesser features,—I would earnestly recommend the frequent and attentive survey of a glacier or glaciers from a considerable elevation above their level and under varying effects of light. Had I confined myself to studying crevasses on the surface of the glacier, measuring their depths, injecting the ice with fluids and taking its temperature; useful and important as these inquiries

---

* See Huber-Burnand's conclusive experiments on this subject, Ann. de Chimie et de Physique, xli. 166, and Fechner's Repertorium, i. 65.

† See the Map of the Mer de Glace and its tributaries in my Travels in the Alps of Savoy.

are, (and I might almost include the fundamental and most important inquiry of all, that of ascertaining the velocity of its parts,) I should have been much longer in seizing the general truth of the individual character of a glacier, the import- ance of the fluid-like connection of its parts, the perfectly secondary importance or unimportance of the fissures by which it is often traversed. The traveller who winds his tortuous and sometimes perilous path amongst these crevasses, forgets, in the fatigue of his circumventions, in the wonder of his curiosity at their beauty and seemingly unfathomable depth, in the appalling steepness of their sides and the com parative insecurity of his own footing—he forgets, I say, in the midst of all these claims upon his attention, his curiosity, and his strength of mind, the comparatively large surfaces of unbroken ice over which he heedlessly walks, and the small, the very small depth at which most of the yawning crevasses which make such an impression on his imagination, dwindle into mere slits;—and when his walk is finished, he imagines that a glacier is a mere network of fissures interlacing in all directions. But let him gain a bold height above its surface, 800 to 1000 feet at least*, so that the whole may be spread somewhat like a map before him, yet not too distant to prevent his seeing the number and forms of the crevasses, and estimating their area compared to that of the unbroken ice, his opinion is first shaken and then changed. He sees in the glacier a *whole*, which, regarded as such, is merely scarred, not dis- sected by these fissures;—he sees a mass as capable at least of conveying strains as thrusts; of which the cohesion is no more destroyed than (to use a comparison which I long ago employed) a parchment sieve is incapable of being stretched, be- cause it is covered with fine slits.

I am confident that this will be plain to every unprejudiced person who will make the observation which I have recommended, and I have no hesitation in stating my belief that it will be found to be fully confirmed by M. WILD's map of the glacier of the Aar, should it ever be published; I say so without having any recollection how the matter stands, although I once had an opportunity of seeing that fine work for a few minutes; and the verification of this remark, by positive measurement, will, so far as I see, be the chief result likely to flow from the patient and disinterested labour of that competent surveyor.

But if this be true in a merely *superficial* plan, how much more true would it be if we could pare off the upper stratum of the glacier, and view a horizontal section of it at a depth of a hundred feet! The depth of the crevasses has, I am persuaded, been as much exaggerated as the thickness of the ice of the glacier has been under- rated. In how few cases (where a glacier does not descend tumultuously) can we

* I may mention, as the very best stations which I am acquainted with, the summit or higher slopes of the hill of Charmoz above Montanvert, Station G*, above Trelaporte, and a point directly above the Couvercle at least 1200 feet higher than the Mer de Glace, which may easily be reached from the glacier of Talefre. Other glaciers offer of course similar points, but few so advantageous ; the glacier of the Aar from the Schnee- bighorn, the lower glacier of Grindelwald from the slopes of the Mettenberg, the glacier of the Rhone from near the Mayenwand, and that of Zermatt from the Riffelberg, are examples.

let a plumb-line down even fifty feet without grazing the sides! and to what an insignificant fissure has the gaping crevasse dwindled even at that small fraction of the glacier's thickness! Supposing the crevasse to become uniformly narrower, how soon would it be extinct!

Again, the crevasses which traverse the surface of the glacier have almost always a determinate direction or directions, of which the simplest type seems to be that of perpendicularity to the veined structure*, which, generally speaking, occasions a convexity of the lines of fissure towards the origin of the glacier. Opposite Montanvert the crevasses form two systems inclined 65° to one another, but this appears to be a casual occurrence arising from a fresh strain being imposed on the ice owing to its rigidity when the direction of the bed or trough suddenly changes, and the two-fold systems probably coexist but for a short space, one tending to close whilst the other opens. Be this as it may, unless where a glacier is falling headlong in the manner of a cascade, the crevasses do not produce any actual dislocation of its mass into blocks or fragments, since the crevasses rarely intersect even where most numerous, but almost invariably *thin out* in the solid mass, whilst another crevasse takes its origin a little to one side or other, leaving a firm connection of ice between them; and the difficulty and danger of traversing a glacier where much crevassed, does not arise from the necessity of leaping from square to square of ice, but from having to traverse these bridges of icy communication, which even there link the glacier together, and which are almost always sharp on their upper edge when the season of the year is pretty far advanced, owing to the continual dripping.

The occurrence of crevasses which cut up a glacier into square or trapezoidal blocks, is sufficiently infrequent to deserve notice. Such occur when a glacier of the second order descends over a boss of granite, or a surface convex in all directions. We have then radiating crevasses combined with concentric ones, producing a tartan-like appearance. Such may be seen in a glacier of the second order on the south side of the Aiguilles of Charmoz and Grepon, above the Glacier du Géant; and it is a very convincing proof of the *essential tenacity* of a glacier, that, with a surface so scarred and intersected, the fragments do not fall away in avalanches. This only is to be explained by the consideration that, thin as are the glaciers of the second order, the apparent dislocation is only superficial.

Were the inequality of the central and lateral movement of the glacier mass to be attributed to longitudinal fissures or discontinuities, by means of which broad stripes of ice slide past each other, we should have to demonstrate the existence of such fissures, which could not be always close unless either (1) the surfaces were mathematically adapted to slide over one another, or (2) the ice possessed sufficient plasticity to mould the surfaces to one another's asperities, in which case the plasticity would alone be sufficient without the discontinuity to explain the motion of the ice. These longitudinal fissures, cutting the common transverse fissures perpendicularly,

* Travels in the Alps, p. 171.

would divide the glacier even where most level into trapezia, and no transverse crevasse could be straight-edged but must be jagged like a saw, or cut *en échelon.* Such a phenomenon never occurs unless where a glacier is moving *torrentially,* or with great disturbance and down a steep. *There* such longitudinal fissures may occasionally be seen, but they form the exception and not the rule. It has been demonstrated by an elaborate proof in § 5, that the only trace of longitudinal discontinuity in the normal condition of the glacier is to be found in the veined structure, which, being caused by a partial discontinuity at a vast number of points, admits of an insensible deformation of the glacial mass without sudden or complete rents, or slips, or the formation of zigzag crevasses.

The existence of the great transverse crevasses, which, even in glaciers not moving torrentially, divide the surface of a glacier by rents perhaps 2000 feet long*, have been thought by some to be comparable to beams of an elastic material, supported at the two ends, and bending under their own weight forward, in the middle. Were this the case, it would scarcely modify the plastic theory as I have propounded it; because in order that such a bar of ice should conform to the known movements of the glacier, opposite the Montanvert for instance, the centre must continually gain upon the sides at the rate of 150 feet per annum at least, consequently the limit of cohesion of an elastic solid would soon be overpassed, and plasticity in the material sufficient to explain the whole motion would inevitably be admitted at last. Independently of this, it is evident, that were such a flexure essential to the motion, the lines of crevasses would be convex in the direction in which the glacier is moving instead of towards its origin.

Argument from the Equable Progression of Glaciers.—The equability of the motions of the various parts of a glacier, united as I have shown them to be by intricate relations†, must, I think, appear conclusive to every one capable of forming a just opinion on the subject, that the relative movements of the various parts of the glacier are due to the action of forces at small distances and to the antagonism of molecular cohesions and molecular strains, and not to the casual jumbling of a quantity of rude fragments. To myself, I confess that this now appears the strongest argument of all for considering the glacier as a united mass like a river, in which there is a nice equilibrium between the force of gravitation, acting by hydrostatic pressure, and the molecular resistances of the semi-solid; the degree of regularity of the law which connects the partial movements is wonderful, and I maintain that it is inexplicable except upon the viscous theory. Thus (1) the glacier moves continually, summer and winter, day and night, and never by fits or starts; for if it does—if gravitation overcomes mere friction, it occasions a shock or avalanche; (2) its mean annual motion is nearly alike from year to year; (3) the relative velocities of points widely distributed over the glacier (but exposed to similar influences of climate), change simultaneously in the same directions, often in the same proportions; thus " the variation of velocity

* Travels, p. 171, 2nd ed.   † See § 5 of this paper, pages 167 and 168.

in the breadth of a glacier is proportional to the absolute velocity at the time of the ice under experiment*." (4) The progression of velocity from the side to the centre is marked by insensible gradations†. (5) When we compare the motion of a given point of a glacier any day of one year and the same day of another, the probability is that the velocity will be exactly the same, if the season be equally hot or cold; hence, surely, a most unexpected result, which I first announced in 1842, that *a few days' observation of a glacier will enable any one to compare its mean rate of motion over its various parts and with different glaciers.* Thus, the motion of a point marked D 2 on the Mer de Glace, was in 1842, from August 1 to August 9, $16\frac{1}{3}$ inches daily; from August 9 to September 16, 18 inches; now next year, 1843, *one* observation at the same point in August gave 16 inches; and in 1844, *one* observation in September gave $17\frac{1}{2}$ inches. But still further, (6) the very law of flexure of the ice is the same from year to year: a series of stations across the ice at the Montanvert gave, in 1842, the following (simultaneous) relative velocities‡ :—

1·000　　　1·332　　　1·356　　　1·367.

The same points being recovered in 1844, the relative motions were (by a single observation of the space moved over in five days)—

1·000　　　1·339　　　1·362　　　1·374,

ratios almost the same but slightly increasing, which corresponds with the fact mentioned above (3), that when the absolute velocities are greater, the relative velocities are so too, which was here the case, for the velocity denoted by 1·000 was a little greater in the second case than in the first.

Tensions and Thrusts.—The occurrence of open crevasses plainly indicates the existence of strains in the ice of glaciers producing disruption, at least partially. Hence some writers have precipitately inferred that the whole glacier must be in a state of tension; an uncertain inference surely in a problem of singular complexity, and one which is not warranted by a more accurate analysis. Yet for a time rival theories seemed poised on the inappropriate question, " Are glaciers in a state of internal tension or compression?" Even if the glacier moved as a mass of fragments, therefore without tension, the cohesion must first have been broken before it could be reduced into fragments. I have been inconsiderately censured for quoting, with approbation§, the observation of M. ELIE DE BEAUMONT, that a glacier appears to be rather in a state of distension than compression, whilst I adopted a hydrostatic pressure, acting from the origin as the source of motion. A careful examination of the passages in question will show that my assent to the view of M. ELIE DE BEAUMONT was limited to *portions* of the glacier, and especially to those portions most crevassed, the parts, namely, which connect the sides and centre, and which serve to drag the more sluggish, because retarded, lateral portions after the freer central part

---

* Travels, p. 148.　　　　　† See § 5 of this paper.

‡ Travels, 1st edit., p. 146.　　　§ Travels, 1st edit., pp. 178, 370; 2nd edit., p. 370 and *note*.

on which the *vis a tergo* acts with most advantage; and in a direction generally parallel to the blue bands, so far as they are due to inequalities of motion in the horizontal plane *. My earliest attempts to obtain clear views of the internal forces acting on a semi-rigid body, impelled by self-contained hydrostatic forces, convinced me how little could be founded on the completeness of any mathematical investigation of them, which in our present state of knowledge may well be considered as hopeless; and reserving to myself the not so difficult task of extricating at a future time the more important practical laws of these strains and thrusts, I very carefully avoided, in my first publication, any allusion to what might be considered as their actual distribution; a distribution varying not only from point to point of the glacier surface, but throughout its thickness, and most undoubtedly varying also for the same point at different seasons of the year, or even changing its sign, so that a tension at one season may become a thrust at another.

I had no reason to repent of this caution, from which I only departed so far in my Seventh Letter on Glaciers, published subsequently, as to deduce in an approximate manner, from elementary mechanical laws, the directions of the *surfaces of tearing* within such a mass as I had described, upon the simple supposition that the hydrostatic pressure acting uniformly, the tendency of motion of any particle will be in the direction of least resistance when all the resistances are taken into account, and that the surfaces of rupture will divide particles whose motions are dissimilar, but will not divide particles whose motions are alike. I repeat that I had no reason to repent of my abstinence from theorizing, when I found that a far better mathematician than myself, taking up the inquiry where I had left it, and after applying himself for a long time to the exclusive mechanical considerations which the viscous theory had suggested, left the subject, as I conceive, little more advanced than he had found it, and fell into some mistakes and inconsistencies, almost inseparable from this way of treating a problem which extensive observation and patient thought can alone disentangle.

Formation of Crevasses.—It has been seen in the third section of this paper, that De Saussure, and almost all his successors, have regarded the crevasses as *accidents* of glacier motion, and not essential to it; and in this view I of course concur. Nevertheless, the study of crevasses is one of considerable, though secondary interest, and is very far indeed from being completed. It requires, among other things, a very sedulous attention to the state of the glacier at various seasons, and even whilst covered with snow; and it requires further a two-fold classification of crevasses, into those which may be considered as proper to the mass of the glacier, and those which merely seam its surface.

I will first speak of the last point.

Though the formation of a crevasse betokens a local distending force, such a force cannot with any certainty be referred to the whole depth of the glacier below the

* See Philosophical Magazine, May 1845, p. 408.

2 D

point where the chasm opens. On the contrary, there is a fully greater probability that under that very spot the ice is compressed. If one cause of a crevasse be, as is universally acknowledged, a protuberance or inequality in the bed over which the ice is impelled, for the same reason that a beam, broken by means of weights, is in a state of longitudinal compression below, where its surface is concave, and of disten-sion above, where its surface is convex, the cracks in the glacier may be due solely to this last and partial cause. Superficial crevasses may consequently be occasioned where there is no *general* distension of the mass, either (1) by the shoving of the semi-rigid glacier as a whole, over a convex declivity, or (2) from an internal tur-gescence arising from hydrostatic pressure, resisted by the intense friction of the ante-rior or more advanced parts of the glacier, which, causing the line of least resistance to be upwards and forwards, forces the pasty mass to tumefy or increase in thickness, exactly as it has been seen in § 2, p. 153, that sluggish lava streams do in a similar case. But if the tumefaction be pushed beyond the limits of plasticity of the superior and more distended portions, they must burst and assume the crevassed forms ac-tually observed in the plastic models described in p. 144. Hence the existence of crevasses not only does not always result from a state of *general* distension in the glacier, but may arise from the precisely contrary condition of great internal compres-sion. This argument is well illustrated by the recent observations of M. Agassiz's co-operators on the glacier of the Aar, whose observations I have elsewhere shown* to be incompatible with any other view than that of intense longitudinal compression in the mass generally, and yet the surface abounds in crevasses of the usual form and dimensions.

The manner of formation of crevasses generally, including such as may betoken a real distending force acting on any part of a glacier throughout its thickness, is not only a most curious question in itself, but suggests others which a correct theory of glacier motion can alone answer. If a crevasse once formed remain a fissure in the ice for ever after, why is the horizontal projection or ground plan of the crevasses of a canal-shaped glacier convex towards the origin of the glacier, and not protuberant in the direction of its motion, as the ascertained greater velocity of the centre would assign? Why are the crevasses for the most part vertical and not inclined forwards, or at least not notably so, on the same account? Why, if the glacier be urged down-wards by a longitudinal force distending it, do not the crevasses continually widen in proportion as they are further from the origin? These questions seem incapable of a sound answer except by supposing that the crevasses are, at least in a great de-gree, the fresh production of every spring, and arise from the sudden start which the glacier makes when that extremity which descends into the valley begins to experi-ence the thawing effects of returning summer. I should not wish to speak positively upon what involves a difficult if not impossible observation,—the state of the glacier with respect to crevasses whilst still under the winter's covering of snow. But the fact

* Ninth Letter on Glaciers. Appendix to Travels, 2nd edit., p. 443

of the transverse direction of the crevasses, or even their convexity towards the origin, from year to year, seems to admit of no other explanation. But besides this, I can affirm, from a careful observation of the crevasses of the Mer de Glace from June to September in one year, that the changes which they underwent were such as preclude the possibility of a crevasse of autumn being merely preserved by the snow of winter, and re-appearing afresh in spring as it had done the previous one. The thing is impossible, because the character of the crevasse is essentially altered. In order that an autumnal crevasse may become a spring crevasse it must be sealed up, annihilated, and opened again. A glance at the three sections in Plate X. fig. 3, will illustrate this. No. 1 shows crevasses freshly opened soon after the snow has quitted the surface of the ice—the edges are sharp, the sides vertical, the openings so small that they may be easily stepped across, and in other instances they are not wider than may admit the blade of a knife. No. 2 shows the crevasse opened to its widest extent by the acceleration of the motion, by the force of the sun which has altogether wasted away the side with the southern exposure, and by the copious drippings of the melting ice and mild rain. No. 3 (which as well as No. 2 is taken from a sketch on the spot, No. 1 being done from recollection) shows what I have elsewhere called the state of collapse of the glacier, which affords the most direct possible evidence of its plastic condition; for we there see, not merely the prominences worn away and blunted by the heat of summer, but subsiding into the hollows, the crevasses being choked by the yielding of their sides, and the glacier again resumes a traversable character, only that the plane surface of spring is changed into irregular undulations preparatory to a complete amalgamation of the whole glacier into one mass *.

The collapse is thus described in my Journal of 1842, written at the time, and therefore more emphatic and unbiassed than after my theoretical views had been matured and published. " 1842, Sept. 16, Friday. The level [of the Mer de Glace at the 'angle'] has sunk since the 9th of August, nine feet 8½ inches. The effect of this immense fall is abundantly evident in this part of the glacier. On my first visit this time [i. e. after an absence of a month], on the 10th, I was quite struck with its shrunk appearance, as I was today with the collapsed state of the crevasses. There cannot be a question but that the glacier has subsided bodily into its bed, and that the semifused pliancy of its materials causes them to recover a uniform and lower level. The crevasses are much less deep than in July and August, as at that time they were larger and more numerous than in June. They are collapsed and (opposite Trelaporte) almost soldered up; the edges all rounded and melted by the sun's heat." The phenomena here described, " the shrunk appearance," " the *semifused pliancy*," " the soldered crevasses," " the rounded edges," convey to the attentive spectator an intuitive conviction of the plasticity of ice at the thawing season, which no words can

* See Travels, p. 174; and Fourth Letter on Glaciers.

2 D 2

express, no mathematical symbols weave into a demonstration.   I can only say that it is easier to believe than to disbelieve; and that sooner or later, it will, I doubt not, be generally admitted.

Considering the crevasses as chiefly superficial in the normal glacier (I mean that of which the inclination is not excessive), it is evident that the formation of the crevasses must depend mainly upon the configuration of the bed.   Where the section of the bed parallel to the length of the glacier is convex upwards, there the tension at the surface will cause the crevasses to expand; when the bed is concave and the surface is being compressed, the crevasses tend to close.   Hence the surface of the glacier descending an irregular bed may be alternately in a state of distension and compression, and the crevasses do not tend to widen indefinitely, which would be the case if the whole glacier were distended.   This tendency in the crevasses to expand and contract in accordance with their position is beautifully seen in viewing the Mer de Glace from a height, as we have recommended.   The steep fall opposite Trelaporte shows the expansion of the crevasses, but the comparative level opposite the little glacier of Charmoz gives it time to recover its solidity by the general closing of the crevasses under compression.   The careful study of such a scene as this gives a more clear insight into the glacier phenomena than any other part of the inquiry, excepting only the measurement of velocities.

Law of Velocities.—To these velocities we now return.   The varying velocities in different glaciers, at different seasons and in different parts of the same season, are all in accordance with the motions of a viscous or plastic body.   They depend upon the slope, being greatest, *ceteris paribus*, when the slope is greatest; and upon the climate to which the glacier is exposed, being greatest in glaciers which descend into deep valleys, and least in those which, though very steep (such as that of the Schönhorn described in § 6), are placed in so elevated and therefore dry and cold an atmosphere as to afford in sufficient water to moisten the snowy mass or *névé*, and which are therefore endowed with very feebly hydrostatic qualities.   This is demonstrated on the one hand by the extreme smallness of their motions, and on the other by the insignificant streams of water to which they give birth even in the height of summer.   In any individual glacier the velocity of the parts must (on any theory) vary with the area of section through which the ice stream has to pass; but yet it may happen that the contraction of a valley, if not accompanied (as is often the case) with an increased slope, will oppose so great a resistance to the efflux of the mass, that under intense longitudinal compression its forward motion is retarded, and the condition of uniform discharge is satisfied by the accumulation of the ice in a vertical direction, the rise of the surface being necessarily accompanied with a thrust from below upwards, and a sliding of the particles over one another in that direction. This appears conclusively to be the case for a great extent of the lower part of the glacier of the Aar, as already mentioned, and affords the most direct evidence which could be desired, that the kind of internal motion necessary for producing the *frontal dip* in the veined structure (which arises from tearing or crushing in sliding in the vertical plane*) was correctly foreseen.

The law of velocities at different points of the axis of a glacier from its origin towards its termination, must evidently depend upon the configuration of each particular glacier. It may be constantly increasing from the origin to the extremity, it may be diminishing, or it may have alternations of increase and diminution; and upon this circumstance the frequency and magnitude of the crevasses will mainly depend. But the *régime* of the glacier, by which we mean to express the combination of circumstances determining its motion, varies from one season of the year to another, owing not only to the general influence of heat and cold, but also to the progressive communication of that influence to portions of the glacier in successive stages of elevation. Evidently the extremity nearest to the valley will receive the earliest and most violent impression of solar heat, whilst the middle and upper regions are involved in complete winter. Partial dilatations must take place in spring, partial condensations in the decline of the year; as is evident from the consideration that temperatures inferior to freezing do not sensibly affect the motion of the ice (see above, p. 192) which higher temperatures do, consequently the influence of season will be chiefly felt in those parts of the glacier where the temperature of the air seldom falls in summer to $32°$, whilst the more stable motion of the higher part acts as a drag or equalizer upon the whole system. The condition of violent distension produces crevasses, that of violent compression produces the frontal dip of the veined structure, or that share of it which is due to the relative motions in a vertical plane. The longitudinal veins will result whether the axis of the glacier be distended or compressed. Hence the reason why the frontal dip is difficultly seen in all the middle region of a glacier, which like the Mer de Glace, is subject to much extension due to great and increasing declivity, and to be well seen must be sought for in the higher parts of the glacier, as above Trelaporte, at the foot of the Couvercle†, and in glaciers subject to great compression, as that of La Brenva, the glacier of the Rhone, the Aar, &c.

Ablation of the Surface.—One phenomenon is most satisfactorily explained by the variations of velocity established and illustrated in this paper. The collapsed state of the glacier after the hot summer of 1842, and the absolute lowering of its surface level by thirty feet in the space of a few months, had struck me as requiring an energy altogether extraordinary in kind and degree to restore next spring the level which had been lost, in order to allow for an equal ablation the succeeding summer; and at first I was disposed to admit so much of the dilatation theory to be true as would account for the swelling of the surface in a vertical direction by the freezing during winter of the infiltrated water‡. Further reflection convinced me however that this explanation was insufficient and also not required, and I accordingly concluded

---

* Seventh Letter on Glaciers. Appendix to Travels, p. 435.

† Travels, 2nd edit., p. 167.   ‡ Fourth Letter. Travels, App., p. 415.

" that the main cause of the restoration of the surface is the diminished fluidity of the glacier in cold weather, which retards (as we know) the motion of all its parts, but especially of those parts which move most rapidly in summer. The disproportion of velocity throughout the length and breadth of the glacier is therefore less; the ice more pressed together and less drawn asunder; the crevasses are consolidated, while the increased friction and viscosity causes the whole to swell, and especially the inferior parts, which are most wasted*." I have nothing to add to this explanation, except that the observation of the motion throughout the whole year confirms it in every particular. The more elevated portions of the glacier, which during a large portion of the year are exposed to a mean temperature under 32°, move in a manner comparatively uniform, the lower extremities undergo great oscillations in their speed (in the ratio of four or five to one; see page 185); hence the attenuation during the summer *régime* which is owing to the drag taking place downwards in an excessive degree; but the winter's cold, equalizing in some measure the velocity everywhere, brings the plasticity into full action, fills the crevasses and swells the surface to its old level.

As it is universally admitted that the glacier proper does not grow in thickness by snowy accumulations, the important variations in its level in different years† cannot be ascribed to the severity of certain seasons increasing the mass of snow falling upon it, but rather to the prolongation of the winter cold into spring and summer, which causes the condensing or accumulating process to be in excess, and therefore the thickness of the plastic mass to accumulate beyond its due amount.

Thus we have the following phenomena, all independently observed, reconciled and explained by one hypothesis; the general convexity of the crevasses upwards, notwithstanding the excess of motion in the centre; the general verticality of the crevasses, notwithstanding the retardation of the bottom; the perfect state of the crevasses every spring succeeding their visible collapse in autumn; the ascertained velocity of different parts of the glacier, and the diversity of the annual changes which these velocities present; the seemingly opposed facts showing the glacier to be subjected to powerful tension, producing crevasses, and yet to be under a compression which produces in some places the *frontal dip*; and finally, the renewal of the level of the ice during winter, which has been lost partly by superficial melting, but as much or more so by the attenuation and collapse of the glacier during summer. These various effects of one cause, though they do not embrace all the phenomena of glaciers, certainly include a very remarkable and complicated group of facts.

* Travels, p. 386, 2nd edit.

† For instance, it has been seen from BALMAT's narrative (p. 189 above), that in 1845 the glacier attained a much higher level at the *Angle* than it had done for three previous years at least, since all the marks of measurements which were cut on the rock in 1842 were concealed: and he attributes this, apparently with reason, to the extreme lateness and coldness of the spring.

Plasticity—Veined Structure.—I certainly never expected, when promulgating the viscous theory, that it would have met with so much opposition on the ground that the more familiar properties of ice are opposed to the admission of its plasticity; and that the fragility of hand specimens should be considered as conclusive against the plastic effect of most intense forces acting on the most stupendous scale upon a body placed in circumstances which subject it to a trial, beneath which the most massive constructions of the pyramid-building ages would sway, totter, and crumble. In an age when generalizations of the more obvious kinds are no longer proofs of genius and perspicacity, and when popular writers on science delight to startle their readers by showing how bodies the most dissimilar possess properties in common; in an age in which *gradations* of properties and organs have been studied with such persevering sagacity, and in which so many unexpected qualities have been discovered;— when iron is classed as a combustible, when metals are found which float on water and which catch fire on touching ice, when a pneumatic vacuum is formed and maintained in vessels five miles long, and whose sides are ripped open twenty times a day; —when, moreover, the simpler abstractions of former times are being daily overset, when no body seems to possess any one property in perfection, and all seem to possess imperfectly every quality admitting of degree; when adamant is rejected from our vocabulary, and softness means only less hardness, and the definition of a perfect fluid is as imaginary as that of a solid without weight;—when a vacuum and a plenum are alike scoffed at, and even the heavenly bodies toil through media more or less resisting; when no substance is admitted to expand uniformly by heat, when glass may be considered a conductor of electricity, and metals as imperfect insulators;—in these days, when the barriers of the categories are so completely beaten down, I had not expected to meet with so determined an opposition to the proposition that the stupendous aggregation of freezing water and thawing ice, called a glacier, subjected to the pressure of thousands of vertical feet of its own substance, might not under these circumstances possess a degree of yielding, moulding, self-adapting power, sufficient to admit of slight changes of figure in long periods of time. Still less could I have anticipated that when the plastic changes of form had been measured and compared, and calculated and mapped, and confirmed by independent observers, that we should still have had men of science appealing to the fragility of an icicle as an unanswerable argument! More philosophical surely was the appeal of the Bishop of Annecy from what we already know to what we may one day learn if willing to be taught: " Quand on agit sur un morceau de glace, qu'on le frappe, on lui trouve une rigidité qui est en opposition directe avec les apparences dont nous venons de parler. *Peut-être que les expériences faites sur de plus grandes masses donneraient d'autres résultats*.*"

---

* Théorie des Glaciers de la Savoie, p. 84. Quoted in my Travels, p. 367, 2nd edit. Since this paper was read, Mr. CHRISTIE, Secretary of the Royal Society, has kindly communicated to me a very striking remark upon a well-known and easily-repeated experiment. The experiment is this. If, in the course of a severe

The " ductility" is indeed not great ; the compact ice even of the slowest moving glaciers bears evidence, in the veined structure, or " blue bands," to the bruise which it has received from the all-powerful strain which has acted on it.   When the difference of motions is excessive, or the slope occasions the speed to be greater than permits the gradual molecular adaptation of the semi-rigid parts to one another, the masses are broken up and fall more or less tumultuously ; *the strain being then removed by the dislocation, the veined or bruised structure is invariably extinguished at last.*   I shall quote a series of examples of the gradation of phenomena, which I conceive to be plainly connected by a common cause.

1. In any torrential glacier, such as the Glacier des Bossons, the upper part of the glaciers of La Brenva, Allalein, or the Rhone, and many others, the fractures are so numerous that the ice descends in blocks, almost as water in a cascade often does in spray, and hence the internal strains being destroyed, no structure is developed, or if previously developed, tends to wear out*.

2. In a glacier moving torrentially, that is with frequent and considerable changes of velocity, but without being divided into blocks by intersecting crevasses, we find real internal cracks in the ice, some feet in length, and an inch or more in thickness, marked by the pure frozen water which fills these spaces in the comparatively opake whitish ice of which glaciers descending rapidly from the region of the *névé* are composed.   Such are peculiarly visible in the lower and more accessible region of the glacier of Bossons† ; perhaps the most instructive which can be named as showing these infiltrated cracks, which by their dimensions, direction, and in every other particular, form a true link between the longitudinal dislocations of a torrential glacier, and the perfect veined structure or bruise into which it passes by imperceptible gradations, including a perfectly regular development of the frontal dip, where we might expect it to be well shown, for the observations of page 184 show that the lowest portion of

winter, a hollow iron shell be filled with water and exposed to the frost with the fuze-hole uppermost, a portion of the water expands in freezing, so as to protrude a cylinder of ice from the fuze-hole ; but if the experiment be continued, the cylinder continues to grow, inch by inch, in proportion as the central nucleus of water freezes.   " In the first instance," says Mr. CHRISTIE, " a shell of ice containing water was formed, no doubt, within the iron shell, and the fuze-hole might be filled by the expansion of the water in the act of freezing ; so that there may be no reason for attributing plasticity to the ice as far as this goes ; but the shell of ice once formed, and the fuze-hole filled with ice, the subsequent rise of the ice must have proceeded from the ice of the interior shell being squeezed through the narrow orifice.   No thawing took place during the process.   Does not this show plasticity even in very small masses of ice ?"   I have also been lately informed, on excellent authority, that in a new work by a most eminent German mineralogist, the plastic character of ice in masses is assumed as an admitted fact.   In corroboration of what has been said in the text, I may farther add, that whilst these sheets are passing through the press, I observe in the Athenæum (June 20, 1846), an account of a patent process for moulding solid tin into tubes and other utensils, in the course of which it is stated that " tin under a pressure of about twenty tons to a circular inch, will *run* according to the law of fluids."

* See Third Letter on Glaciers.   Travels, Appendix, p. 407.

† See Travels, p. 181.

the glacier of Bossons moves slower than its middle portion; there is therefore a manifest longitudinal compression arising from the friction of the bed*.

3. The next stage is that of the perfect bruise or veined structure, best seen in the most united and least fissured parts of glaciers with rocky sides and moving over a moderate slope. Whatever increases lateral compression (without however necessitating dislocation), such as the union of two or more glaciers in one, tends to develope the structure more perfectly†. Such cases are well seen on several parts of the Mer de Glace, and of the glacier of Miage‡.

4. In very wide glaciers, moving with feeble velocities, the veined structure is slightly developed, except near the sides, simply because the *twist* being small the ice is hardly bruised. Nor can we wonder to find the structure at the distance of many hundred yards from the sides of a vast slow-moving glacier of this description, if developed at all, to be complex and irregular, exhibiting twists such as I have figured in my Travels, p. 164, and which are peculiarly conspicuous in the magnificent glacier of Aletsch. This circumstance finds a precise analogue in the case of a great river, such as the Rhine, or indeed in any river moving with a very slight inclination; the excess of velocity of the central above the lateral parts, not very great at any rate, is distributed over such a space that the slightest casual disturbance of the current, from an irregularity in the bottom or sinuosities of the course, produces local differences of velocity, occasioning ripples and eddies in various parts of the breadth. If these ripples and eddies, in other words, differential motions of adjacent particles, could be visibly represented by using differently coloured fluids, they would undoubtedly afford sections exhibiting undulations and contortions exactly like those which the ice presents in the cases mentioned above. We claim therefore the apparent exception as a real proof of our general rule.

5. In the *névé* proper, no true veined structure is developed; *first*, because, whilst the mass is snowy, its powdery nature yields without admitting of a fracture or bruise; *secondly*, because the true *névé* has rarely any lateral compression worth mentioning, being widely spread and not contained between steep barriers; *thirdly*, because its motion is altogether very small; *lastly*, because its extreme dryness does not afford water enough to percolate its substance and there to be frozen; when it does so, it ceases to be *névé*.

On these grounds I hope that the theory of the veined structure, so important to that of glaciers, may be considered as explaining a number of intimately connected phenomena.

* The internal rents in the lava of Zafarana referred to in § 2 of this paper, and figured in Plate IV. fig. 8, present a perfect analogy with those of the glacier of Bossons, and appear to be due to the same cause.

† Third Letter. Travels, Appendix, p. 407.

‡ See the figures of the structure of the glacier of Miage. Travels, p. 197.

"The glacier struggles between a condition of fluidity and rigidity*." " A glacier is not a mass of solid ice, but a compound of ice and water more or less yielding, according to its state of wetness or infiltration†." "The pressure communicated from one portion to the other, will not be the whole pressure of a vertical column of the material equal in height to the difference of level of the parts of the fluid considered; the consistency or mutual support of the parts opposes a certain resistance to the pressure, and prevents its indefinite transmission. * * A glacier is not coherent ice, but a granular compound of ice and water‡." " When the semifluid ice inclines to solidity during a frost, the motion is checked; if its fluidity is increased by a thaw, the motion is instantly accelerated. * * It is greater in hot weather than in cold, because the sun's heat affords water to saturate the crevasses§." Such were the terms in which within a few months after suggesting the viscous theory I expressed my opinion of the influence of the compound structure of the glacier, a mass composed, not of ice alone, but of ice including water in its countless capillaries never frozen‖ even in winter. The quality of plasticity or viscosity resulting from the union of a nearly perfect fluid with an imperfect solid is seen in very numerous and familiar instances, as for instance in sand, which is itself devoid of any tenacity until its interstices have been saturated with just so much water as to cause it to flow; or in the still more familiar instance of water-ice prepared for the table, in which the varying proportion of the solid and fluid ingredient gives to it every shade of consistency, from a brittle solid to a liquor including suspended solid grains. The prodigious effect of capillary infiltration in determining the motion of even the most solid and ponderous bodies, breaking up their parts, and giving to the motion of the whole a more or less river-like character, is seen in the frequent case of land-slips, as for instance that of Goldau. And scarcely less instructive are the numerous examples, cited in the first section of this paper, of huge masses of almost cold and brittle lavas being pressed on with a uniform and graduated motion, by the almost unimpeded hydrostatical communication of pressure from the yet active fluid which circulates unseen in their pores. With this analogy before me, I replied in 1844 in the following terms to the question, " How far a glacier is to be regarded as a plastic mass?" "Were a glacier composed of a solid crystalline cake of ice, fitted or moulded to the mountain bed which it occupies like a lake tranquilly frozen, it would seem impossible to admit such a flexibility or yielding of parts as should permit any comparison to a fluid or semi-fluid body transmitting pressure horizontally, and whose parts might change their mutual position so that one part should be pushed out whilst another remained behind. But we know in point of fact, that a glacier is a body very differently constituted. It is clearly proved by the experiments of AGASSIZ and others, that the glacier is not a mass of ice, but of ice and water; the latter percolating freely through the crevices

---

* Third Letter on Glaciers, August 1842. Appendix to Travels, p. 407.

† Travels, 1st edit., 1843, p. 175.

‡ Travels, p. 367., edit. 1843.   § Ibid. p. 372.   ‖ Ibid. p. 361, 372.

of the former to all depths of the glacier; and as it is matter of ocular demonstration, that these crevices, though very minute, communicate freely with one another to great distances, the water with which they are filled communicates force also to great distances, and exercises a tremendous hydrostatic pressure to move onwards in the direction in which gravity urges it, the vast porous crackling mass of seemingly rigid ice in which it is, as it were, bound up*."

Now the water in the crevices does not constitute the glacier, but only the principal vehicle of the force which acts on it, and the slow irresistible energy with which the icy mass moves onwards from hour to hour with a continuous march, bespeaks of itself the presence of a fluid pressure. But if the ice were not in some degree ductile or plastic, this pressure could never produce any, the least, forward motion of the mass. The pressure in the capillaries of the glacier can only tend to separate one particle from another, and thus produce tensions and compressions, *within the body of the glacier itself*, which yields, owing to its slightly ductile nature, in the direction of least resistance, retaining its continuity, or recovering it by re-attachment after its parts have suffered a bruise, according to the violence of the action to which it has been exposed.

The action of warm weather in accelerating the movement of the glacier is plainly due to the abundance of the water saturating its pores; but this may act in two ways; first, by rendering the frame-work of ice less brittle when it is in the very act of dissolving by the circulation of water in a perfectly fluid state through its pores†, and secondly, and more particularly, from the hydrostatic effect of *gorging* a porous mass with fluid. When an incipient frost dries even momentarily the surface of the glacier, the vast porous mass begins to *drain*. This is a very slow process, owing to the resistance to the passage of a fluid through very long and complicated canals. Were it not so, glaciers would be entirely dry after sunset and in winter, which is not the case. The hydrostatic pressure within the whole glacier is however sensibly diminished by the process of drainage; this is evident from watching the level of water in a vertical hole of any depth made within the solid ice of the glacier. After much rain or heat this level is always higher than after dry cold. In the former case the glacier may be said to be gorged, the supply of water from the surface exceeding the power of the drainage to carry it off. The circulating vessels are therefore overcharged. In the latter case the superficial supply is stopped, the drainage goes on slowly though uninterruptedly, and the level of the water in the vertical shaft slowly descends, indicating the diminution of internal pressure. If it were not for the capillarity of the ducts, it is plain that no effective hydrostatic pressure would

---

* Sixth Letter. Appendix to Travels, 2nd edit., p. 428.

† This I think is undeniable, from the appearance of the collapsed crevasses above referred to, notwithstanding the difficulty of imagining any variation in the sensible heat of water circulating in ice. It is not the only fact in the glacier theory which seems to require some modification of the commonly received laws of latent heat at the very limit of congelation and liquefaction.

be developed at all; the flow being equal to the supply, no part of the *vis viva* would be expended in producing internal pressures.   With this concluding observation I commit the Plastic or Viscous Theory of Glaciers to the impartial judgment of those qualified to decide on its merits in explaining facts, and on the variety of difficult and complicated considerations which opposed and still oppose themselves to a complete development of it.

Edinburgh, Jan. 10, 1845.

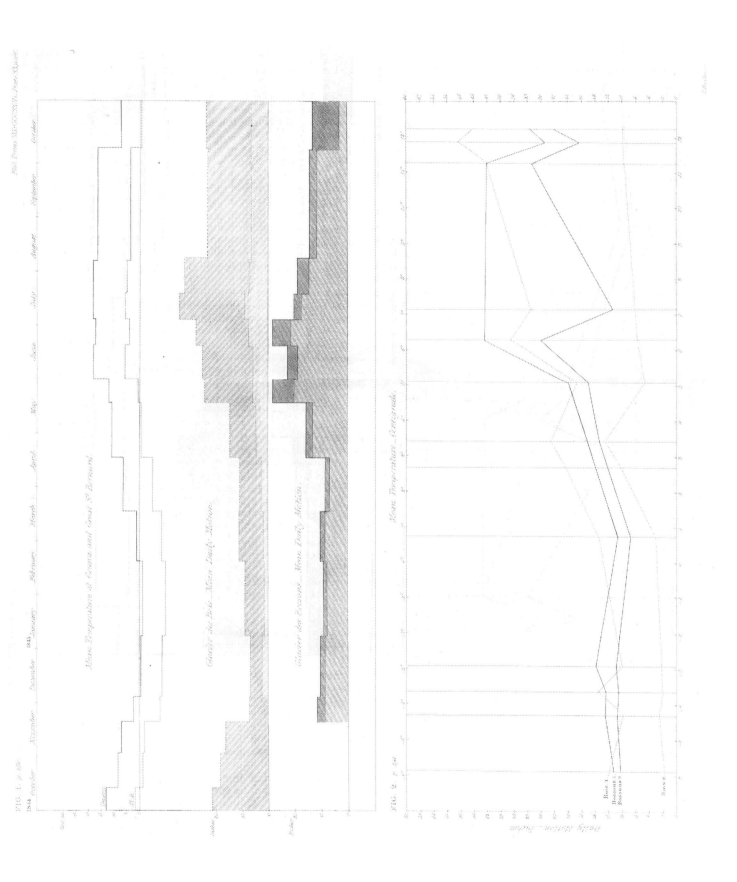

Fig. 1. *p.178.*

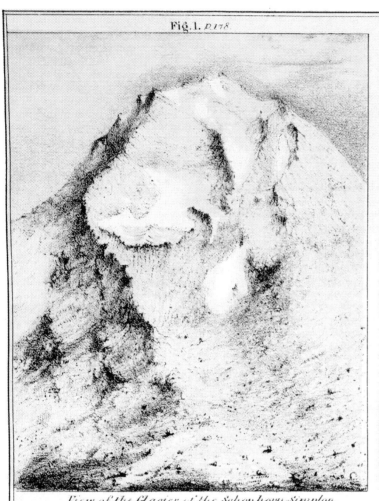

View of the Glacier of the Schönhorn-Simplon.

Fig. 3. *p.201.*

Nº 1. SPRING.

North   *South*

Nº 2. SUMMER.

Nº 3. AUTUMN

Fig. 2. *p.178.*

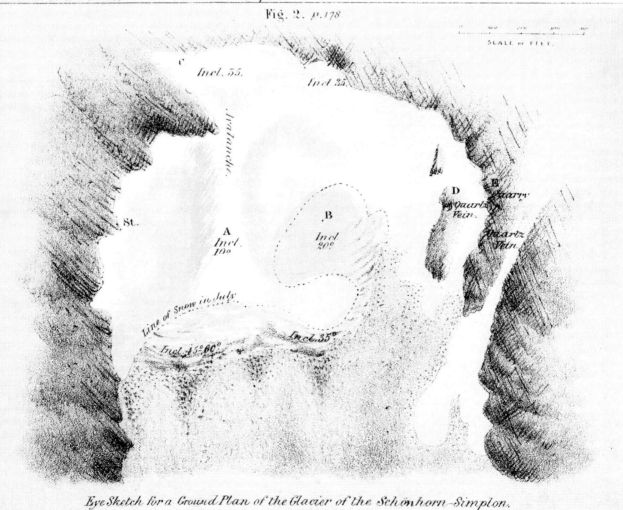

SCALE of FEET.

Eye Sketch for a Ground Plan of the Glacier of the Schönhorn-Simplon.

XIV. *On the Physical Phenomena of Glaciers.*—Part I. *Observations on the Mer de Glace. By* John Tyndall, *F.R.S., Membre de la Société Hollandaise des Sciences; la Société de Physique et d'Histoire Naturelle de Genève; la Société Philomathique de Paris; Mitglied der Naturforschenden Gesellschaft, Zürich, der Gesellschaft zur Beförderung der Gesammt-Naturwissenschaften, Marburg, der K. Leop. Akad. der Naturforscher, Breslau, and Professor of Natural Philosophy at the Royal Institution.*

Received May 20,—Read May 20, 1858.

§ 1.

The Philosophical Transactions for 1857 contain a paper by Mr. Huxley and myself upon the Structure and Motion of Glaciers. The observations on which that paper was founded extended over a very brief period, and hence arose the desire, on my part, to make a second expedition to the Alps, in which I regret to say my friend was unable fully to join. The phenomena of the Mer de Glace being those on which the most important theoretic views of the constitution and motion of glaciers are based, I wished especially to make myself acquainted by personal observation with these phenomena. Six weeks of the summer of 1857 were accordingly devoted to the examination of this glacier. For the purpose of observing its motion, bearings and inclinations, and also of determining its width at various points, I took with me an excellent 5-inch theodolite, and a surveyor's chain; for both of which I am indebted to the kindness of the Director-General of the Geological Survey, and to Professor Ramsay. I propose to divide the investigation into two parts, the first of which forms the subject of the following paper, while the second will be the subject of a future communication. It gives me great pleasure here to record my grateful sense of the able and unremitting assistance rendered me throughout the entire period of the observations, by my friend Mr. T. A. Hirst, whose name indeed, had he permitted it, I should gladly have seen associated with my own at the head of this paper.

§ 2. *On the Motion of the Mer de Glace.*

Our first observation of the motion of the Mer de Glace was made on the 14th of July. On the steep terminal incline of the Glacier de Bois we singled out a tall pinnacle of ice, the front edge of which was perfectly vertical. In coincidence with this edge I fixed the vertical wire of our theodolite, and after three hours found that the ice cliff had moved downwards, the cross hairs being now projected against the face of the cliff several inches above its edge.

Our first line across the glacier was set out upon the 17th of July. The mode of proceeding in all such cases was this:—the theodolite was placed beside the glacier, quite clear of the ice, and usually at a sufficient height above it to command an uninterrupted view across the glacier. The plummet of the instrument being suspended, a stake was driven into the ground, or a fixed stone was carefully marked exactly under the point of the plummet. The direction of a line perpendicular to the axis of the glacier from this point being ascertained, a well-defined object was sought in the production of this line, at the opposite side of the valley—the sharp edge of a cliff, a projecting corner of rock, or a well-defined mark on the surface of the rock. This mark, and the objects surround-

Fig. 1.

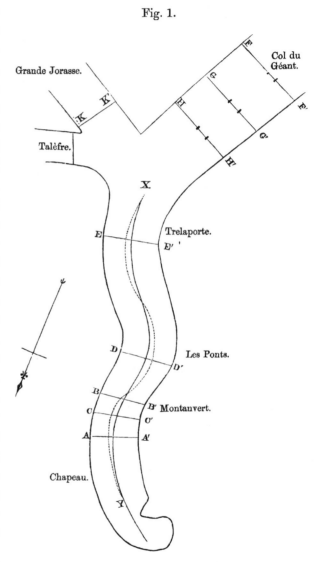

ing it, were carefully sketched, so that in coming subsequently to the place the line was immediately recognized. The cross hairs being fixed upon the mark, the object-end of the telescope was lowered until the cross hairs cut the point at which a stake was to be placed. The positions of the stakes were found by means of an ordinary traveller's *baton*, which was set erect upon the ice and moved up or down in accordance with the signals from the observer at the theodolite, till the exact point was hit upon. Here the ice was pierced to the depth of about 18 inches, and a wooden stake was firmly driven into it. The position of each individual stake was secured by taking the angle of depression down to it, a precaution which was found very useful when subsequent reference to any particular stake was necessary. The exact time at which each stake was driven in was noted; and the time at which the displacements were measured being also observed, the motion was afterwards reduced, by calculation, to its diurnal rate.

The station from which our first line started was at some distance below the Montanvert Hotel, and about eighteen yards, in an ascending direction, from the station marked D on the Map of Professor FORBES*. The line is that marked AA′ on the sketch-map, fig. 1†.

* We found this station marked by a chisel on a block of granite, and painted red.

† The side of the glacier opposite to the Montanvert is much crevassed, and while fixing a stake upon one

On the 18th of July we set out a second line above the Montanvert Hotel, and we afterwards measured the displacements of the stakes along the line AA'. The result led to the establishment of a hitherto unobserved law of glacier motion, which the discussion of the observations will gradually render manifest. Reduced to twenty-four hours, the motion of the stakes along our first line was as follows:—

First Line (AA').—Mean Daily Motion.

No. of stake.	Motion in inches.	No. of stake.	Motion in inches.
West 1	$12\frac{1}{4}$	6	
2	$16\frac{3}{4}$	7	$26\frac{1}{4}$
3	$22\frac{1}{2}$	8	
4	$25\frac{1}{2}$	9	$28\frac{3}{4}$
5	$24\frac{1}{2}$	East 10	$35\frac{1}{2}$

Stake No. 7 of this series was about midway between the bounding sides of the Mer de Glace; No. 1 was near the lateral moraine at the Montanvert side, and the retarding influence of this side is very manifest. With slight breaches of regularity, the rate of motion increases gradually from the first stake towards the centre of the glacier.

But it will be observed that stake No. 7 by no means moves the fastest. Stake No. 10 stood far beyond the centre, and upon the portion of the glacier derived from the Léchaud and Talèfre. This portion is distinguishable at a glance by the quantity of dirt upon its surface, the portion derived from the Glacier du Géant remaining comparatively clean throughout the entire length of the Mer de Glace. Professor FORBES accounts for the excessive crevassing of the eastern side of the glacier by assuming that the Glacier du Géant, having by far the greater mass, moves most swiftly, drags its more sluggish companions after it, and thus tears them asunder. The foregoing observations show that this assumption is untenable. The difference here observed cannot be referred to the slip to which reference has already been made in the note at the foot of this page, for the slip did not amount to more than 4 inches at the utmost. Further, the displacements were measured a second time on the following day, when the maximum movement of the Glacier du Géant portion was found to be $27\frac{1}{2}$ inches, and that of the Léchaud and Telèfre side $32\frac{1}{2}$.

Our second line, marked BB' upon the sketch-map, had its terminal station on the ancient moraine a little higher up the glacier than the Montanvert Hotel. Along this line thirty-one stakes were driven on the 18th of July, and their displacements measured the day following. The results reduced to twenty-four hours are as follows:—

of the ice ridges here, the whole mass slid suddenly some inches forward. Were special attention directed to the crevassed portions of a glacier, the same phenomenon might, I doubt not, be frequently observed.

Second Line (BB′).—Mean Daily Motion.

No. of stake.	Motion in inches.	No. of stake.	Motion in inches.
West 1	$7\frac{1}{2}$	17	$22\frac{1}{2}$
2	$10\frac{3}{4}$	18	21
3	$12\frac{1}{4}$	19	$22\frac{1}{2}$
4	$14\frac{1}{2}$	20	$20\frac{1}{2}$
5	$14\frac{1}{2}$	21	×
6	16	22	×
7	$16\frac{3}{4}$	23	$24\frac{1}{2}$
8	$17\frac{1}{2}$	24	×
9	19	25	$21\frac{3}{4}$
10	$19\frac{1}{2}$	26	×
11	$19\frac{1}{2}$	27	×
12	21	28	$22\frac{1}{4}$
13	21	29	$22\frac{3}{4}$
14	21	30	$25\frac{1}{4}$
15	$22\frac{1}{2}$	East 31	$25\frac{3}{4}$
16	$22\frac{1}{2}$		

The stakes marked thus × were fixed by the eye, their positions being such that they could not be seen by the theodolite. Some of them were placed in deep glacial hollows, where, without an instrument, it was difficult to keep them in the same vertical plane. The slight uncertainty thus arising induced me finally to reject them. The gradual augmentation of velocity from the side towards the centre is very manifest; but it will be observed that stake 31, which stood upon the Talèfre side of the glacier, moved quickest of all. The difference in favour of the latter side is, however, much less than it was lower down.

The reason why in the two cases just considered the terminal stake towards the eastern side of the glacier shows no retardation, is, that the state of the ice, and the position of the theodolite, were not such as to enable us to continue the line of stakes completely across the glacier to the eastern side, and hence the observations could not show the retarding influence of that side. In setting out the third line CC′, therefore, Mr. HIRST took up a position on the Chapeau side of the valley, from which the vision across the glacier was quite uninterrupted by ridges or other obstacles, while the crevasses were not impracticable. One of the fixed termini of this line was the corner of a window of the Montanvert Hotel. There were twelve stakes planted along the line, and the motion of these during twenty-four hours, from the 20th to the 21st of July, was as follows:—

Third Line (CC').—Mean Daily Motion.

	East.										West.	
No. of stakes.	1	2	3	4	5	6	7	8	9	10	11	12
Motion....	$19\frac{1}{2}$	$22\frac{3}{4}$	$28\frac{3}{4}$	$30\frac{1}{4}$	$33\frac{3}{4}$	$28\frac{1}{4}$	$24\frac{1}{2}$	25	25	18	×	$8\frac{1}{2}$

Stake No. 1 was fixed in the ice, close to the eastern side of the glacier, and the retarding influence of this side is quite manifest from the measurements. A glance, however, reveals a fact confirmative of the former measurements; the daily motion of the extreme eastern stake is $14\frac{1}{2}$ inches behind the maximum, while the motion of the extreme western stake is $25\frac{1}{2}$ inches behind it. The stake No. 5, which moved at the maximum rate, was also much nearer to the eastern than to the western side of the ice-stream; the observation therefore corroborates those already made as regards the position of the point of maximum motion.

How then is the fact to be accounted for, that the point of maximum motion of the Mer de Glace is thus thrown towards its eastern boundary? Reflection suggested to me that the effect might be due to the curvature of the valley through which the Mer de Glace moves. At the place where the foregoing observations were made the glacier bends, turning its concave side to the Montanvert, and its convexity towards the Chapeau. M. RENDU insists on the complete analogy of the phenomena of a river and those of a glacier; and the idea has been to a great extent corroborated by the measurements of Professor FORBES and M. AGASSIZ; but let us make a bolder application of the analogy than any of them contemplated, confining our view to the influence of curvature merely. The point of maximum motion of a river moving through a channel similar to that occupied by the Mer de Glace, would lie on that side of the centre of the channel towards which the river turns its convex curvature. Can this be the case with the ice? If so, the place of maximum motion ought to be different where the glacier bends in the opposite direction. Fortunately the Mer de Glace itself enables us to bring this idea to a test.

Higher up the valley, and opposite to the passages called "Les Ponts," such a band occurs. Here the convexity is turned towards the Montanvert or western side of the valley. A line was set out across this portion of the glacier on the 25th of July, and its measurement upon the 26th gave the following results:—

Fourth Line (DD').—Mean Daily Motion.

	East.													West.		
No. of Stakes.	1	2	3	4	5	6	9	10	11	12	13	14	15	16	17	
Motion..	$6\frac{1}{2}$	8	$12\frac{1}{2}$	$15\frac{1}{4}$	$15\frac{1}{2}$	$18\frac{3}{4}$	$19\frac{1}{2}$	21	$20\frac{1}{2}$	$23\frac{1}{4}$	$23\frac{1}{4}$	21	$22\frac{1}{4}$	$17\frac{1}{4}$	15	

After the setting out of this line, its length was measured by Mr. HIRST; and found to be 39 chains 25 links, which, as each chain is equal to 22 yards, gives 863 yards as the width of the Mer de Glace opposite the first "*Pont.*" A mark on the rock crossed by this *pont* constituted indeed one of the fixed termini of the line.

For the sake of stricter discussion, a copy of the notes of this measurement faces the next page.

The stakes along the line are marked thus, $\odot$. The fixing of them commenced at the Echellets or eastern side of the valley, and they were numbered *from* this side: the measurement, on the contrary, commenced at the "Pont." Hence it is that the 17th stake was the first encountered in the measurement. This stake stood at a distance of 326 links, or nearly 72 yards from the edge of the glacier. Stake No. 1 at the other end of the line stood close beside the lateral moraine at the eastern side of the glacier.

Referring to the notes, it will be seen that the place of maximum movement occurs between the stakes 12 and 13, the former at a distance of 1544 links, and the latter at a distance of 1250 links from the western side of the glacier. The mean of these is 1397 links; consequently, as the entire width is 3925 links, the point of maximum motion is here 1131 links nearer to the western than to the eastern side of the Mer de Glace. The dirt also which marks the junction of the portion of the ice derived from the Col du Géant, with that derived from the other tributaries, is crossed at the distance 2251; hence the place of maximum motion occurs at a point 854 links *west of the dirt*, while on the lines set out lower down the point of maximum motion was far in upon the dirt, eastward from the junction. The position of the point of maximum motion changes, therefore, in exact accordance with the explanation given above.

But the question is capable of still closer examination. The notes enable us to compare a number of points at the eastern side of the glacier with others, situated at the same respective distances from the western side. Let us call every pair of points, one of which is situated as far from the eastern boundary as the other is from the western, *corresponding points*. The corresponding points along our fourth line may then be ranged as follows:—

$$
\begin{array}{lccccc}
 & \text{S. V.} & \text{S. V.} & \text{S. V.} & \text{S. V.} & \text{S. V.} \\
\text{West} \ldots & 17-15 \; ; & 16-17\tfrac{1}{4} ; & 15-22\tfrac{1}{4} ; & 13-23\tfrac{3}{4} ; & 12-23\tfrac{1}{4} \\
\text{East} \ldots & 3-12\tfrac{1}{2} ; & 4-15\tfrac{1}{4} ; & 5-15\tfrac{1}{2} ; & 7-18\tfrac{1}{4} ; & 9-19\tfrac{1}{2}
\end{array}
\Big\} \ldots (A)
$$

The numbers under the letter S are those of the stakes, those under V are the corresponding velocities. It will be seen that in each case the point on the western portion of the glacier moves quicker than the corresponding point on the eastern side. As a whole, therefore, the western side moves more speedily than the eastern, which is the reverse of what was observed lower down, but quite demonstrative of the explanation which refers the effect to the curvature of the valley.

An inspection of the notes also shows, that at the place where the fourth line crossed the glacier, the crevasses are found chiefly upon the portion derived from the Glacier du Géant. The dirt which announces the position of the other tributaries of the Mer de Glace is crossed at the distance 2251; and after this distance we find the remark "crevasse nearly closed," "closed crevasse;" so that not only is the eastern side of the glacier here less crevassed than the western, but crevasses previously formed are partially, or wholly closed up. The shifting of the place of strain consequent on the change of curvature, carried naturally along with it the shifting of the crevasses. It may be inferred from the notes that the measurement of such a line is not without its difficulties.

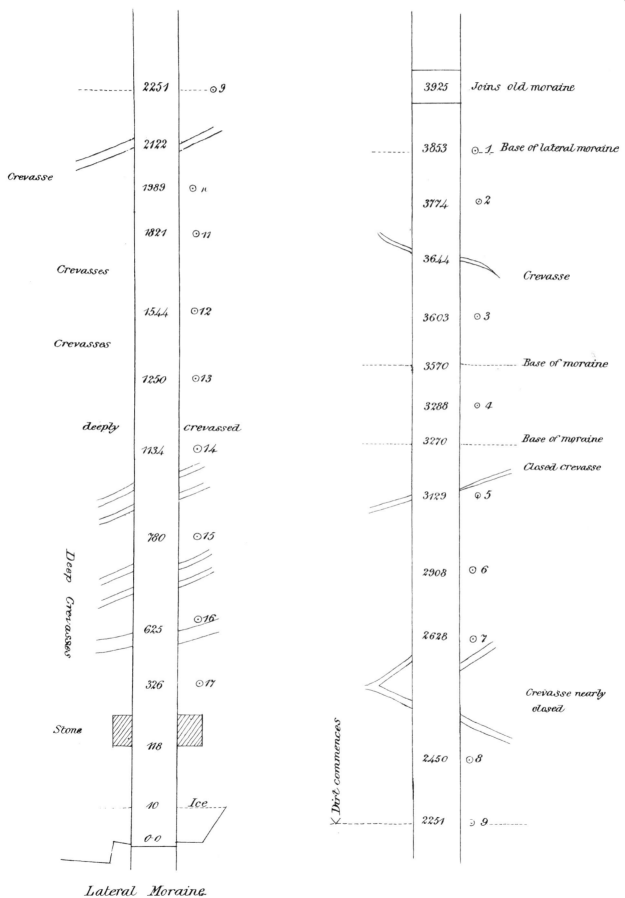

Lateral Moraine.

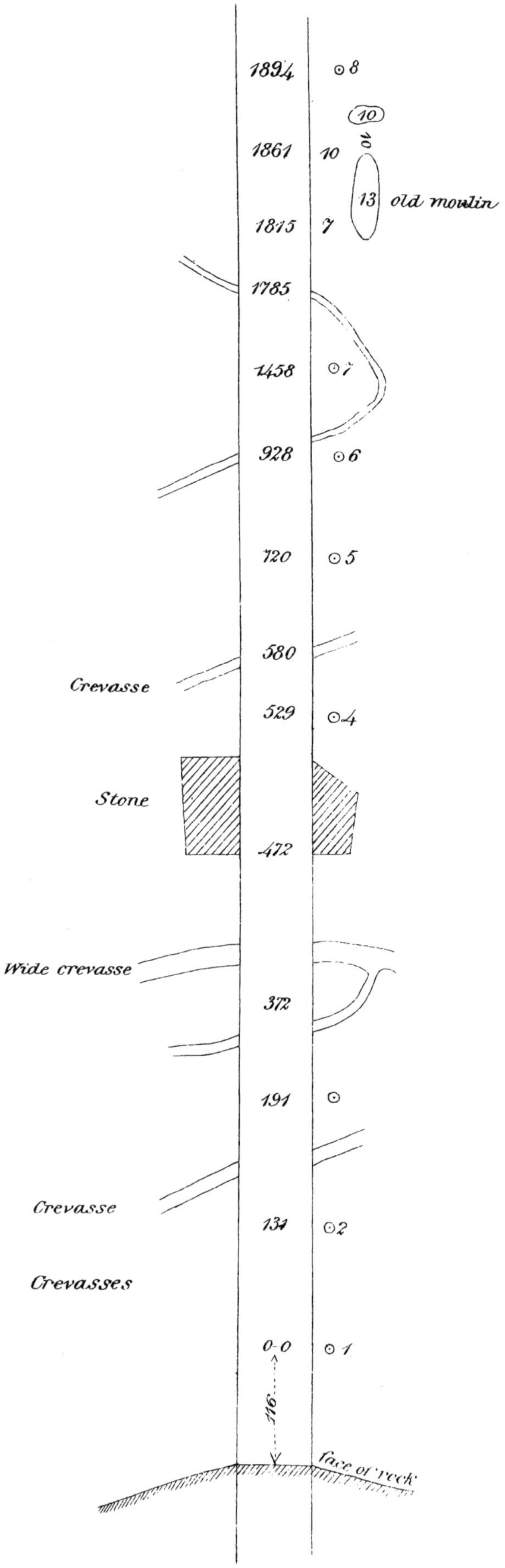

1894 ⊙ 8

(10)

1861 10 10

13 old moulin

1815 7

1785

1458 ⊙ 7

928 ⊙ 6

720 ⊙ 5

580

Crevasse

529 ⊙ 4

Stone

472

Wide crevasse

372

191 ⊙

Crevasse

134 ⊙ 2

Crevasses

0-0 ⊙ 1

116

face of rock

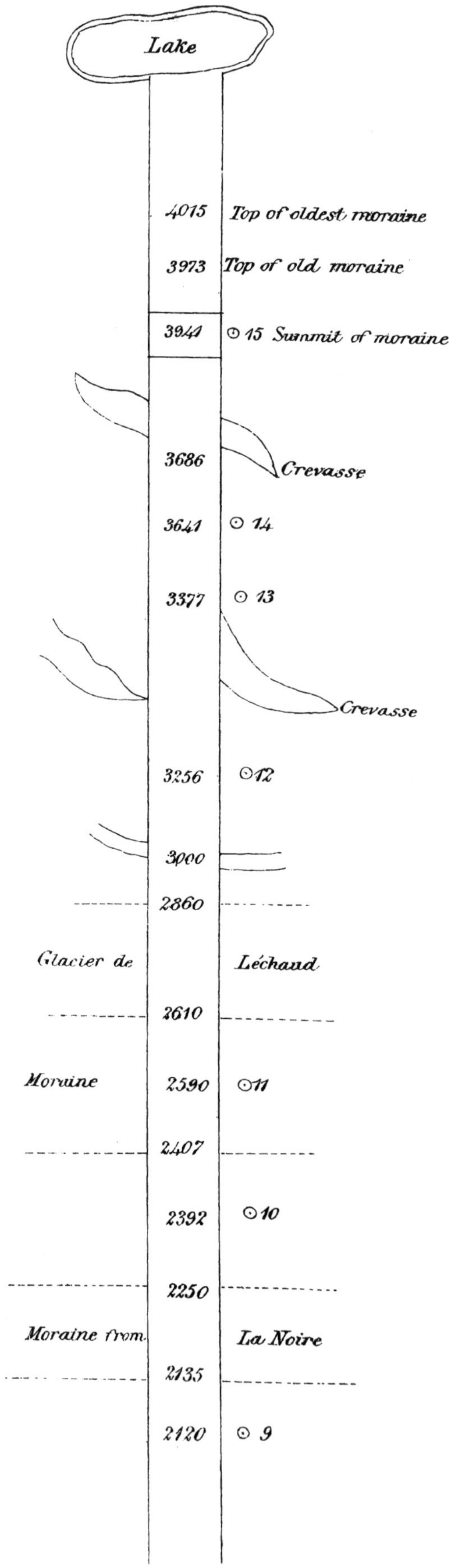

Lake

4015 Top of oldest moraine

3973 Top of old moraine

3944 ⊙ 15 Summit of moraine

3686 Crevasse

3641 ⊙ 14

3377 ⊙ 13

Crevasse

3256 ⊙ 12

3000

2860

Glacier de Léchand

2610

Moraine 2590 ⊙ 11

2407

2392 ⊙ 10

2250

Moraine from La Noire

2135

2120 ⊙ 9

Our next line (EE') stretched across the glacier from the promontory of Trelaporte to the base of the Aiguille du Moine. The instrument being placed upon a grassy slope above the promontory, the line was set out on the 28th of July. The Trelaporte end of this line was immediately under the station marked G* on the Map of Professor FORBES; the displacements of the stakes were measured on the 31st of July, and were found to be as follows:—

Fifth Line (EE').—Mean Daily Motion.

West.														East.
No. of Stakes.	1	2	3	4	5	6	7	8	9	10	11	12	13	14	15

Motion...... $11\frac{1}{4}$ $13\frac{1}{2}$ $12\frac{3}{4}$ 15 $15\frac{1}{4}$ 16 $17\frac{1}{4}$ $19\frac{1}{4}$ $19\frac{3}{4}$ 19 $19\frac{1}{2}$ $17\frac{1}{2}$ 16 $14\frac{3}{4}$ 10

The first of these stakes was about 80 feet distant from the face of the rock at Trela-porte; the 15th was on the lateral moraine, which moved along with the ice at the opposite side of the valley. The retarding influence of both sides is very clearly shown, the motion of the central stakes being nearly twice that of the extreme ones. As a whole, the rate of motion is slower here than at the " Ponts" or at the Montanvert.

This line was also chained by Mr. HIRST; a copy of his notes, showing the distances along the line at which the stakes were set, faces this page.

The chaining commenced at a point 116 links distant from the face of the rock at Trelaporte. Adding these 116 to the distance 3941, we have 4057 links, or 893 yards for the width of this portion of the Mer de Glace. The point of maximum motion occurs at stake No. 9, which is 2236 links distant from the rock at Trelaporte, or more than one-half the distance across; that is to say, the point of maximum motion is here nearer to the Talèfre side than to the Géant side of the glacier. Here, again, we have a result different from that obtained with our fourth line; and if we look to the sketch-map we shall see the reason. Between the fourth and fifth lines the Mer de Glace has passed a point of contrary flexure; and here at Trelaporte the convex side of the glacier is turned towards the base of the Aiguille du Moine.

Taking the 116 links at the commencement into account, the following pairs of stakes may be regarded as corresponding points:—3 and 14; 4 and 12; 7 and 10; the small numbers referring to stakes at the western, and the large numbers to stakes at the eastern side of the glacier. The relative motions of these points are as follows:—

West...... 3—$12\frac{3}{4}$;   4—15;   7—$17\frac{1}{4}$.
East ...... 14—$14\frac{3}{4}$;   12—$17\frac{1}{2}$;   10—19.

Comparing this Table with Table A, we observe a reverse result; in the latter case the western stakes moved most swiftly; here the eastern ones do so; the deportment of the ice is the same as at the places intersected by our three first lines, and the curvature of the valley is also similar.

From the foregoing observations the following law of glacier motion is derived:—
When a glacier moves through a sinuous valley, the locus of the point of maximum motion does not coincide with a line drawn along the centre of the glacier, but always lies on the

convex side of the central line. It is therefore a curve more deeply sinuous than the valley itself, and crosses the axis of the glacier at each point of contrary flexure*.

Fig. 2.

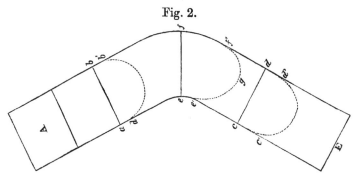

The law may be illustrated by the following experiment:—A, fig. 2, is a box filled with fine mud, which by raising a sluice in front flowed into the curved trough AE. A line *ab* was drawn upon the mud above the bend, a second line, *cd*, below the bend, and a third, *ef*, at the bend. The distortions of these three lines by the motion of the mud downward will reveal the position of the point of maximum motion at the particular places where they are drawn. The line *ab* was distorted to *a'b'*, the summit of the curve being exactly in the centre of the trough, thus proving that the centre was the place of maximum motion. The same was true of the line *cd*, which was distorted to *c'd'*. The line *ef* was distorted to *e'f'*, the summit *g* of the curve being nearer to the side *bfd* of the trough, this proving the point of maximum motion to lie towards that side.

I scarcely know a case more calculated to impress the mind both with the yielding power of ice to pressure, and the magnitude of the forces brought into play in the motion of glaciers, than the crushing of the three tributaries of the Mer de Glace through the throat of the valley between Trelaporte and the base of the Aiguille du Moine. Not wishing to trust the eyes in the estimation of distances here, each of the three confluent branches was measured. The width of the Glacier du Géant, a short distance above the Tacul, was found to be 5155 links, or 1134 yards. The width of the Glacier de Léchaud, just before its junction with the Talèfre, was found to be 3725 links, or 825 yards. That of the Talèfre, before it is influenced by the pressure of the Léchaud, that is, across the ice-cascade, was found, approximately, to be 2900 links, or 638 yards. Adding all together, we find the sum of the widths of the three branch glaciers to be 2597 yards. At Trelaporte these three branches *are forced through a gorge* 893 *yards wide*; and our measurements show that it passes through with a velocity of nearly 20 inches a day!

Limiting our view to one of the glaciers thus compressed, the facts appear still more astonishing. Previous to its junction with the Talèfre, the Glacier de Léchaud has a width of thirty-seven chains and a half. In passing through the jaws of the granite vice at Trelaporte, this broad ice river is squeezed to a driblet *less than four chains in width*! This fact illustrates the relation of the size and power of a glacier to the quan-

* If the defined line between X and Y on the sketch map represents a line drawn along the centre of the glacier, the dotted line will represent the locus of the point of maximum motion.

tity of snow drainage which supplies it.   The Talèfre has its basin, and the Géant has its vast plateau, from which the respective glaciers derive nutrition; but the Léchaud is fed by two or three couloirs merely, which descend principally from the Mont Mallet and Les Jorasses.   The Géant, in the struggle for place at Trelaporte, takes up more than half the valley, and the others come in the order of the drainage which supplies them.

The velocity of the Mer de Glace at Trelaporte being about 20 inches, it seemed probable that the velocity of the Glacier du Géant above the Tacul, and also of the Léchaud above its junction with the Talèfre, would be considerably less, in consequence of the greater width at these places.   This proved to be the case.   On the 29th of July a line was set out across the Glacier du Géant, a little above the Tacul.   There were ten stakes in this line, and their motions reduced to twenty-four hours were as follows:—

Sixth Line (HH') —Mean Daily Motion.

No. of Stakes.	1	2	3	4	5	6	7	8	9	10
Motion	11	10	12	13	12	$12\frac{3}{4}$	$10\frac{1}{2}$	10	9	5

The velocity here is considerably under that of the Mer de Glace at Trelaporte.

On the 1st of August we set out a line across the Glacier de Léchaud immediately above where it is joined by the Talèfre.   The line commenced at the side of the glacier beneath the block of stone called the Pierre de Béranger, and ran perpendicular to the axis of the glacier to the other side.   The displacements were measured on the 3rd of August: reduced to twenty-four hours, they are as follows:—

Seventh Line (KK').—Mean Daily Motion.

No. of Stakes.	1	2	3	4	5	6	7	8	9	10
Motion	$4\frac{1}{2}$	$8\frac{1}{4}$	$9\frac{1}{2}$	9	$8\frac{1}{2}$	$7\frac{1}{2}$	$6\frac{1}{4}$	$8\frac{1}{2}$	7	$5\frac{1}{2}$

The stakes 8 and 9 were at opposite sides of a "moulin," which was found to share the general motion of the glacier.   A new crevasse crossed our line above 8 and below 9, and the greater advance of stake No. 8 was probably owing to the yielding which this crevasse permitted.   The rates of motion, it will be observed, are still less than those upon the Glacier du Géant.

Were the Glacier de Léchaud subjected to no waste during its descent, and did no accumulation take place at any point, equal quantities of ice would pass through all its cross sections in the same time.   The compression which takes place at Trelaporte is not a change of *volume* but of *form*.   The mass is squeezed laterally, and no doubt expands vertically.   Comparing the velocities and widths at Trelaporte and opposite the Pierre de Béranger, we should be led to the result that the depth of the Glacier de Léchaud at the former place would, if no waste had taken place, be at least four and a half times its depth at the latter.   The loss of ice by superficial and subglacial melting must materially modify this result; but some interesting observations might be made in con-

2 o

nexion with the point, and I think one result of such observations would be the establishment of the comparative shallowness of the Glacier de Léchaud.

There is another characteristic of glacier motion which was predicted by Professor FORBES, before any observations had been made upon the point, and afterwards confirmed both by his own measurements and those of M. MARTINS,—I allude to the fact that the glacier is not only retarded by its sides, but by its bottom, the superficial ice thus moving more quickly than that in contact with the bed of the glacier.

Objections have been made to both the measurements alluded to, and I was therefore desirous to submit the question to a new test. The experiments which I have to record were made upon the face of an ice precipice, which offered a rare opportunity for an observation of the kind. The face formed the eastern boundary of the Glacier du Géant near the Tacul, was about 140 feet in height, and nearly vertical. I requested Mr. HIRST to place two stakes, one at the top and the other at the bottom of this precipice. This was done on the 3rd of August; and on the 5th it was found that the stake at the top had moved through 12½ inches, while that at the bottom showed an advance of 6 inches only. There was some uncertainty regarding this latter result, on account of the danger incurred by the assistant, from the stones which fell incessantly from the top of the precipice, and which compelled him to retreat several times before the measurement could be effected.

I was reluctant, however, to leave an observation of the kind with a shade of uncertainty attached to it. On the 11th of August, therefore, I fixed myself two stakes, one at the top and the other at the bottom of the precipice, and feeling strongly impressed with the importance of ascertaining the motion of a point midway between top and bottom, I cut steps in the ice, climbed the face of the precipice, pierced the ice with an auger, and drove a stake firmly into it. Until Monday the 17th of August I was unable to reach the place again. On this day I penetrated through dense fog and snow to the Tacul, and found the highest of the three stakes standing, but the two lower ones were buried in a heap of snow which lay at the base of the precipice. On the following day the perilous process above described had to be repeated; and on Tuesday the 20th of August the displacements were measured. Reduced to twenty-four hours, the motion of the three stakes was found to be as follows:—

		inches.
Top stake	. . . . . .	6·00
Middle stake	. . . . .	4·59
Bottom stake	. . . . .	2·56

The distance from the top of the ice-wall to its base was found, by measurement with a rope, to be 140·58 feet, but it was not quite perpendicular at its upper portion; the height of the middle stake from the ground was 35 feet, and of the bottom one 4 feet. It is therefore proved by these measurements that the bottom of the ice-wall at the Tacul moves with less than half the velocity of the summit; while the deportment of the intermediate stake shows how the velocity increases from the bottom upwards.

§ 3. *On the Cause of Glacier Motion.*

The various theories which have been advanced to account for the progression of glaciers are too well known to need detailed discussion here. SAUSSURE, and some before him, thought that the glacier slid along its bed*. CHARPENTIER thought that the motion was due to the freezing of water in capillary fissures, and the consequent swelling of the contents of these fissures. Other hypotheses have been advanced without producing any deep impression. It has been objected to SAUSSURE's theory, that were it true, glaciers must slide down with an accelerated motion; but reflection alone would deprive this objection of weight, and an experiment of Mr. HOPKINS completely refutes it†. When incessantly checked by the surface over which they slide, even avalanches may, and do, sometimes descend with a uniform motion. The motion of a man in walking down stairs is on the whole uniform, but it is actually made up of an aggregate of small motions, each of which is accelerated. It is easy to conceive that ice moving over an uneven bed, will, when it is released from one opposing obstacle, be checked by another, and its motion thus be rendered sensibly uniform. So many obstacles exist along the bed of a glacier, that sudden slipping forwards of the mass through any considerable distance is not to be expected. But the real weak point of SAUSSURE's theory, though partly true, is its inability to account for many facts observed since his time. The theory of CHARPENTIER, though not always fairly represented, has been shown to be untenable.

The facts submitted to our consideration are briefly as follows:—We see the glacier winding through a valley, squeezing itself through a gorge, and widening where it has room. We see that the centre moves more quickly than the sides, and the top more quickly than the bottom; and the next demand of the mind is for a general principle which shall unite these facts, and from which they shall follow as physical corollaries. Professor FORBES seeks this principle in the *viscosity* of the ice. Ice, according to him, is a substance resembling treacle, honey, or tar, and the observed phenomena are a consequence of this property. In this assumption consists what is called the *viscous theory*‡.

* I hardly think, however, that SAUSSURE would have subscribed to some of the interpretations of his theory now extant.

† See HOPKINS in Philosophical Magazine, vol. xxvi. p. 4. Were it not that this objection is thoughtlessly repeated in every work upon glaciers, I would not dwell upon it here. The objection drawn from the deportment of secondary glaciers lying on steep slopes is also very commonly dwelt upon, but it is equally without weight; and applies with at least as much force to the viscous theory as to the theory of SAUSSURE.

‡ The name of M. RENDU will always be honourably associated with the theory of glacier motion. He first drew attention to the power of the glacier to move through a sinuous valley, to narrow and widen and behave like lava or like "a soft paste." He conjectured also that the centre would move more quickly than the sides. In fact he appears to have had a correct conception of almost all that the subsequent observations of Professor FORBES established. I regret to say that I have not been able to obtain M. RENDU's original memoir.

August 1859.—Thanks to my Zürich friends, I have recently had the pleasure of reading M. RENDU's paper, the perusal of which has confirmed my estimate of his sagacity. Had this gentleman been a philosopher instead of an ecclesiastic, we should doubtless have heard more about his claims than we have hitherto done.

Before entering upon the examination of this theory, I would ask permission to make the following remarks:—I am aware that the paper published by Mr. HUXLEY and myself has produced considerable diversity of opinion among scientific men. Some, whose opinions are entitled to every respect, regard the views there advocated, and the experiments there described, as consistent with and explanatory of the viscous theory; while others, of equal eminence, believe that if the views referred to be sound, the viscous theory can no longer be maintained. Under these circumstances it behoves me to state distinctly the point of view from which I intend to examine the theory, submitting myself completely to the public sense as to whether this point of view be the correct one or not. Both the terms and the illustrations made use of by Professor FORBES have diffused ideas regarding the physical qualities of ice which render a strict examination of the subject essential. Let me here briefly state what I understand by viscosity, and what I, and other more competent persons, at one time believed to be a demonstrated property of ice *.

By viscosity, I understand that property of a semifluid body which permits of its being drawn out when subjected to a force of *tension*, the particles of the substance taking up new positions of equilibrium, so that when relieved from the strain the substance has no distortion to recover from. A capacity to change the form under crushing *pressure* is not, I think, a test of viscosity; for this power is possessed by substances, to which we should never think of applying the term viscous.

In examining whether glaciers possess the power of yielding to tension like viscous bodies, I would refer:—1. To the shifting of the place of strain by the curvature of the valley, to which I have already referred. Let ABCD, fig. 3, embrace a curved portion of a glacial valley, and let AB be a linear element of the glacier transverse to its axis.

Fig. 3.

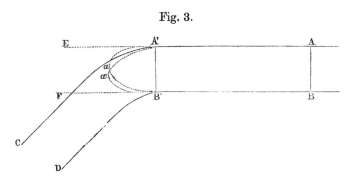

As the ice descends AB becomes curved in consequence of the quicker movement of its centre. Did the valley continue straight in the direction of E and F, the point of maximum velocity would, after a certain time, be found at a, midway between the lines AE, BF; but the curving of the valley throws the point a to a', and thus increases the strain upon the branch $a'A'$ of the curve, while it diminishes the strain upon $a'B'$. The conse-

* "*Gluey tenacity*" is the quality which I have heard ascribed to ice by intelligent and cultivated persons.

quence of this difference of action upon the two branches, is that the side of the glacier which is subjected to the augmented tension does not yield to the strain as a viscous body would do, but *breaks*. In the words of Professor FORBES, the glacier at this place becomes " *excessively crevassed*." This fact, therefore, as far as it goes, is opposed to the idea of viscosity as above defined.

2. The fact that the centre of a glacier moves more quickly than the sides, is that on which the viscous theory is chiefly based: let us examine the circumstances connected with this motion, availing ourselves while doing so both of the figure and the reasoning of Mr. HOPKINS. Let ABCD, fig. 4, be a sloping canal, into which is poured a

Fig. 4.

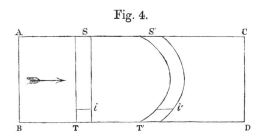

quantity of treacle, honey, tar, or melted caoutchouc, all of which have been referred to as illustrative of the character of ice; and let the mass move down the slope in the direction of the arrow. Let ST be a narrow segment of the viscous substance; this segment, as it moves downwards, will take the form S'T'. Supposing Ti to be a square element of the mass, it will be distorted lower down into the lozenge T'i', and the line Ti will become T'i'. Now the analogy between such a substance and ice fails in this respect; in the viscous mass the short diagonal of the square *stretches* to the long one of the lozenge, but, in the glacier, the ice breaks at right angles to the tension. and *marginal crevasses* are formed. It was by means of the simple diagram here sketched that Mr. HOPKINS showed why the marginal crevasses of a glacier are inclined towards its source*. This fact, therefore, so far as it goes, is also opposed to the idea of viscosity.

But it is known that in the case of a substance confessedly viscous, a sudden shock or strain may produce fracture. Professor FORBES justly urges, " that sealing wax at moderate atmospheric temperatures, will mould itself (with time) to the most delicate inequalities of the surface on which it rests . . . . . but may, at the same time, be shivered to atoms by a blow with a hammer†." Hence, in order to estimate the weight of the objection, that the glacier breaks when subjected to strain, we must know the conditions under which the force is applied.

The fifteenth station on the line (EE') at Trelaporte stands on the lateral moraine of the glacier; between it and the fourteenth, a distance of 300 links, or 190 feet, intervenes. and within this distance the glacier suffers its maximum strain. Let AB (fig. 5) be the

* Philosophical Magazine, vol. xxvi. page 160.

† Philosophical Magazine, Fourth Series, vol. x. p. 201. Proceedings of the Royal Society, June 14. 1855.

side of the glacier, and let the direction of motion be that indicated by the arrow.   Let *abcd* be a square element of the glacier with a side of 190 feet.   The whole square moves downwards with the glacier, but the side *bd* moves quickest.   The point *a* moves 10 inches, the point *b*, 14·75 inches in twenty-four hours, the differential motion thus amounting to an inch in five hours.   Let *ab'cd'* be the shape of the figure after five hours' motion, the distance *bb'=dd'* being 1 inch; then the line *ab* would be extended to *ab'*, and the line *cd* to *cd'*.

Fig. 5.

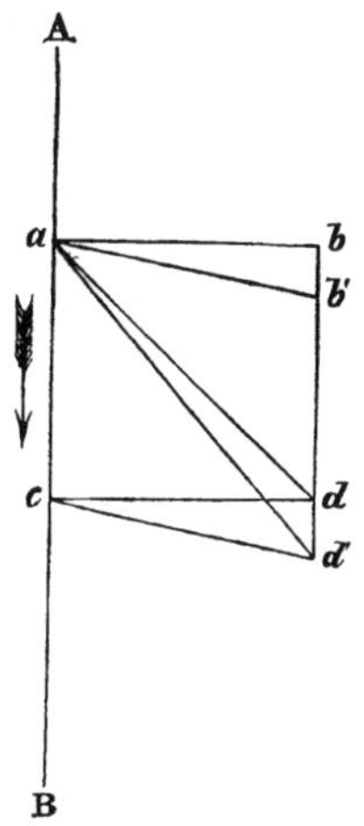

But the extension of *these* lines does not mark the *maximum strain* to which the ice is subjected.   Mr. HOPKINS has shown this strain to take place along the line *ad*, which encloses an angle of 45° with the side of the glacier.   In five hours, then, this line, if capable of yielding, would be stretched to *ad'*.

In the right-angled triangle *abd'* we have *ab*=2280 inches, *bd'*=2281, and hence we find *ad'* to be 3225·1 inches; the diagonal *ad* is 3224·4 inches; and the amount of yielding required from the ice is that the latter line shall be extended by five hours' gradual strain to the length of the former.

This is the utmost demand made upon the presumed viscosity of the ice, but the substance is unable to respond to it: instead of stretching, it breaks, and copious fissures are the consequence.   It must not be forgotten that the evidence here adduced merely proves what ice *cannot* do; what it *can* do in the way of viscous yielding we do not know.   There is no experiment on record, with masses great or small, to show that the substance possesses, in any measurable degree, that power of being drawn out which is the very essence of viscosity.

Further, the case here referred to is not solitary, but typical.   I dare say every single glacier of the first order would furnish proofs of the absence of viscosity equally cogent with that here brought forward.   The marginal crevasses of glaciers usually result from an incapacity on the part of the ice to respond to a demand upon its viscosity, not greater than that just cited *.

When a person unaccustomed to glacier life observes, from a safe distance, the profound fissures by which the ice is intersected, the question sometimes arises, " what if one of these chasms should suddenly open beneath the traveller's feet?"   There is, however, no fear of this.   The crevasses, when first formed, are exceedingly narrow, and they

---

* It may, however, be urged that I do not know how much the ice observed in the locality referred to had been stretched before it arrived there.   Extend an elastic string to the point of breaking, and a small additional force would break it; but this latter small extension would be no measure of the extensibility of the string.   To this I reply, that it is the very essence of a viscous mass to accommodate itself to the forces which act upon it, so that in each new position the texture of the substance shall be in a state of equilibrium.   If such a mass be broken it will have no distortion to recover from.   The idea that a glacier is typified by such a string as that referred to, has been expressly rejected by the ablest advocates of the viscous theory; in proof of which I would refer to the lucid note of Dr. WHEWELL, in the 26th volume of the Philosophical Magazine, page 172.   Cases may occur where the lateral yielding produced by the *pressure* along *bc*, fig. 5, may satisfy the *strain* along *ad*; in such a case no marginal crevasses would be formed.

open with extreme slowness.   While standing one evening, in company with Mr. HIRST, on the Glacier du Géant, both of us were startled by a sound like a heavy explosion in the body of the glacier, underneath the place where we stood.   This was instantly followed by a succession of loud cracks, accompanied by a low singing noise.   The ice continued cracking for an hour; but notwithstanding the manifest breaking of the glacier, which was to some extent awe-inspiring, we could not, for a long time, detect any trace of rupture.   The escape of air-bubbles from the surface first informed us of the position and direction of the incipient crevasse, for such it was.   It was so narrow that the thinnest blade of my penknife would not enter it.

On another occasion, our guide, while engaged in setting out one of our lines, observed the ice to break beneath his feet, and a rent to propagate itself suddenly, with loud cracking, to a distance of 50 or 60 yards across the glacier.   These fissures are produced by tension, and the velocity with which they widen is a measure of the amount of relief demanded by the glacier.   The crevasse last alluded to required several days to attain a width of 3 inches, and the opening of the one on the Glacier du Géant was far slower than this.   This is their general character.   They form *suddenly* and open *slowly*, and both facts are demonstrative of the non-viscosity of the ice.   *For were the substance capable of stretching, even at the small rate at which they widen, there would be no necessity for their formation* *.

There is another point of view from which the question of viscosity may be examined; but as the observations which bear upon it possess a general value, I will devote a special section to them; choosing afterwards those which more particularly apply to the case now under consideration.

§ 4. *On the Inclinations of the Mer de Glace.*

By calculation from heights and distances, Professor FORBES obtained approximately the inclinations of some portions of the Mer de Glace†, but no direct observations on the subject have been hitherto made.   On the 4th of August we transported our theodolite to the Jardin, for the purpose of ascertaining its inclination, and that of the Glacier du Talèfre.   From the green space on which visitors to the place usually repose, the angle of elevation to the top of the Jardin is 24° 7′, and from the same place downwards to the bottom of the Jardin the inclination is 30°.   From the bottom of the Jardin, for some distance along its medial moraine, the ice is nearly level, its inclination being only 21′.   A succession of slopes then follows, enclosing with the horizon the following angles of depression:—3° 5′; 4° 25′; 6° 50′; 8° 5′ and 9° 40′, which last brings us to the brow of the ice cascade.   The inclination of the fall is 25°,—producing a line drawn along the centre of the cascade until it cuts the moraine between the Talèfre and Léchaud: the inclination along this line, from the base of the cascade downwards, is 7° 30′.

* For an interesting account of the formation of a number of new crevasses, see AGASSIZ, 'Système Glaciaire,' p. 310.

† Travels, p. 117.

The descent of the ice through this gorge from the basin of the Talèfre, is adduced by Professor FORBES as an illustration "which will appear to the impartial reader almost a demonstration" of the principle of viscosity. "The ice is compact," he urges, "and almost without fissures. . . . . The open crevasses which commence a little above AB are turned towards the basin*." The line AB here referred to is actually in the jaws of the gorge, and apparently at a considerable distance below where the ice enters it. The description certainly would not apply to the ice of the year 1857. Long before reaching the summit of the fall the most skilful iceman would find himself in difficulties. We proceeded as far as we dared amid the pits and chasms into which the glacier is torn, and which followed each other so speedily, that the ridges between the fissures were often reduced to mere plates and wedges, which were in many instances bent and broken by the lateral pressure. At some places vortical forces seemed to have acted upon the mass, and turned huge pyramids so far round as to place the structural veins at right angles to their normal position. Looking downwards towards the summit of the cascade, the ice was frightfully riven. The glacier descends the cascade itself in wedges, pyramids, and columns, which latter often fall with a sound like thunder, and crush to pieces the ice crags below them. After this description I do not think that the case is likely to be accepted as a demonstration of the viscosity of ice.

I now pass on to the inclinations of the Glacier du Géant. For some distance below the base of the so-called *Seracs* the irregularities of the glacier render an estimate of its general inclination somewhat difficult, but I should judge it to be about 13°. From the end of this steeper portion, two slopes, one of 4° 37', and the other of 3°, bring us to the Tacul, and from this point to the bottom of the ice valley at Trelaporte we have the following series of inclinations:—2° 15'; 3° 15'; 5° and 9°; thence to the Grand Moulin the slope is 3° 30', and afterwards, down the glacier to a point nearly opposite to the Grande Cheminée below l'Angle, the inclinations are 3° 10'; 5°; 6° 25', and 4°. The glacier then descends a slope of 9°, and afterwards passes the Montanvert at an inclination of 4° 45'. Below the Montanvert it falls steeply for some distance, the inclination being 16°. Between the base of this slope and the brow which marks the termination of the Mer de Glace and the commencement of the Glacier des Bois, the slope is 5° 10'. The ice afterwards descends an incline of 22° 20' in a state of great dislocation. From the base of this incline the general inclination of the lower portion of the glacier is 10°.

A brief reference to the Glacier de Léchaud will complete this portion of our subject. The upper portion of the glacier, to the base of the steep snow slopes which rear themselves against the Grande Jorasse, has an inclination of 4° 29'. Opposite to the icefall of the Talèfre, the inclination, for a short distance, is 3° 17', and afterwards down to the Tacul, where the Léchaud and Géant join, the slope is 5° 22'.

I will now endeavour to show the theoretic significance of the observations above recorded, referring in the first place to the great terminal slope of the Glacier des Bois,

* "Reply to HOPKINS," Philosophical Magazine, 1845, vol. xxvi. p. 415.

down which the ice is shot in crags, pinnacles, wedges, and castellated masses, all tossed together in the utmost confusion. Regarding this portion of the glacier, Professor FORBES writes as follows:—"Escaping from the rocky defile between the promontory of the Montanvert and the base of the Aiguille de Dru, it pours in a cascade of icy fragments, assuming the most fantastic forms, into the valley beneath." Above the fall the ice is compact: Professor FORBES compares it to the dark unruffled swell of swift water rushing to precipitate itself in a mass of foam over a precipice.

In fig. 6 I have protracted the inclination of the fall and of the glacier above it, one

Fig. 6.

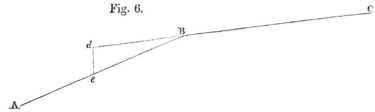

of them, BC, making an angle of 5° 10′, and the other, BA, an angle of 22° 20′ with the horizon. Supposing the ice to pursue the direction which it had previous to reaching the fall, it would, at the end of a certain time, reach the point d *; but the ice is not rigid enough to do this, and the mass descends to e. Now if it be the viscosity of the substance which has carried it in a certain time from B to d, that same property ought, one would think, to enable it to drop down the vertical de without breaking. But so far from its being able to do this, the glacier descends the slope BA as "a cascade of icy fragments." The fact, therefore, adds its evidence to that already adduced against the viscosity of the substance.

But the case will appear much stronger when we revert to other slopes upon the Mer de Glace. For example; the inclination of the glacier above l'Angle is 4°: it subsequently descends a slope of 9° 25′, and in doing so is so much fissured as to be absolutely impassable. The chasms cut the glacier from side to side, and present clear vertical faces of great depth†. Subtracting the smaller of the above angles from the larger, the difference, 5° 25′, gives the *change* of slope which produces the chasms. In fig. 7 the two adjacent slopes are protracted to a proper scale. Now the velocity of the

Fig. 7.

glacier here, in the direction of its length, is to the vertical velocity with which it would have to sink to reach its bed, as Bd : de, or as the cosine of 5° 25′ is to its sine, or as 996 : 94, or, in round numbers, as 10 : 1. Hence if it be viscosity which enables the mass to move from B to d in a certain time, the same property ought, one would think, to permit it to *sink* through the space de, which is only one-tenth of Bd, in the same

* I here assume that the general inclination of the surface of the glacier changes in accordance with that of its bed, which will hardly be questioned.

† I once found myself alone upon this portion of the glacier towards the close of a day's work, and experienced great difficulty in escaping from the entanglement of chasms in which I had involved myself.

time. But this is not the case. In accommodating itself to the change of inclination, the glacier breaks and is fissured in the manner described.

The change of inclination last mentioned, so far from marking the limit at which transverse crevasses begin to be formed, is sufficient to produce chasms of great magnitude, and in most inconvenient numbers. Higher up the glacier, transverse crevasses are produced by a change of inclination from 3° 10′ to 5°. If this change be accurately protracted, the mere inspection of it will illustrate more forcibly than words can do the absence of the power of viscous yielding on the part of the ice.

Looked at broadly, then, two classes of facts address themselves to the attention of the glacier investigator; one entirely in accordance with the idea of viscosity, and the other as entirely opposed to it. The affirmers and deniers of the viscous theory have perhaps been influenced too exclusively by one or the other of these classes of phenomena. The analysis of the facts gives the result, that where *pressure* comes into play we have the evidences of apparent viscosity*, but where *tension* is active we have evidences of an opposite kind. One of these classes of effects is as undeniable as the other, and hence the true theory of glaciers must render an account of both.

When the mountain snow is first moistened, it becomes more coarsely granular; these granules abut against each other, and hold air and water in their interstices. But as successive layers press upon the mass, the granules are squeezed more closely together; rupture and liquefaction, succeeded by regelation, take place at the points of abutment; water and air are expressed by the process, and the mass becomes more and more consolidated. But although powerfully squeezed, each portion of the deeper ice is surrounded on all sides by a resistent mass; it is thus compelled to yield very gradually to the pressure and moves slowly through into the valley of *écoulement*. As far as external appearances go, there is, of course, almost a perfect similarity between such an action and one due to viscosity.

But when a force of tension is applied, the case is wholly different. That intestine mobility which characterizes a truly viscous body, and enables one molecule to move round another while clinging to it, or one particle to advance while another slides in laterally to supply its place, being absent, the only way in which such a body can meet the requirements of a strain is by breaking, the fissures widening as the strain continues.

Thus, I think, we take account of all the facts adduced in proof of viscosity, and also furnish a satisfactory explanation of the other set of facts on which the opponents of the viscous theory have hitherto based their arguments.

Royal Institution, May 1858.

---

* The ingenious experiment of Mr. Christie with a bomb-shell filled with water and submitted to a freezing temperature, belongs, of course, to this class of effects.

XV. *On the Veined Structure of Glaciers; with observations upon White Ice-seams, Air-bubbles and Dirt-bands, and remarks upon Glacier Theories.* By JOHN TYNDALL, *F.R.S., Professor of Natural Philosophy, Royal Institution.*

Received February 24,—Read February 24, 1859.

§ 1. *Introduction.*

ON the 20th of May, 1858, I communicated a paper to the Royal Society, containing an account of observations made upon the Mer de Glace of Chamouni. In addition to the questions there discussed, another of great importance occupied my attention during my sojourn at the Montanvert, and that was the *veined structure* of the ice. To obtain information on this head, I visited almost every portion of the Mer de Glace and its tributaries; I examined the Talèfre and Léchaud glaciers, and spent several days amid the *seracs* of the Glacier du Géant. To investigate the connexion, if any, between the structure of the glacier and the stratification of its névé I ascended the Col du Géant, and afterwards inspected the magnificent ice-sections exhibited in the dislocations of the Grand Plateau and other portions of Mont Blanc.

During this investigation my convictions were by no means fixed; cases strongly suggestive of the influence of pressure, in producing the structure, came before me, and again other cases appeared which suggested, with almost equal force, the influence of stratification. The result, however, of the observations on the Mer de Glace was a strong opinion that *pressure* was the true cause of the phenomenon.

But I could not help feeling that the facts and arguments which I was in a position to bring forward would still leave the question an open one. They might influence the opinions of others, as they had influenced mine; but I had nothing to advance on which the mind could rest with perfect certainty. In short, neither the Mer de Glace nor its tributaries furnished facts capable of completely deciding the question. The subject being one on which a great deal had been written and retracted, I was unwilling to swell the bulk of the literature connected with it, while a possibility remained that what I had to say upon the subject might also require withdrawal. I therefore thought it better to wait another year; to extend the range of my observations, to visit glaciers in which the mechanical conditions of strain and pressure were different from those of the Mer de Glace. Thus by varying the circumstances, and observing Nature at work under different conditions, I hoped to confer upon the investigation the character and precision of an *experimental* inquiry.

The course of the inquiry in 1858 was as follows:—I first examined the glaciers of Grindelwald; crossing the Strahleck, I ascended the lower glacier of the Aar to the

Grimsel, thence to the glacier of the Rhone, thence to the great Aletsch glacier, in the neighbourhood of which I remained eight days. I afterwards spent eleven days at the Riffelberg, and explored the entire system of glaciers between the Monte Rosa and Mont Cervin. I thence proceeded to the Matmark Alp, and remained for five days in the vicinity of the Allalein glacier; I afterwards visited the Fée glacier, and completed the expedition by a visit to the Mer de Glace and its tributaries, and a second ascent to the summit of Mont Blanc.

The present paper contains the evidence derived from the sources thus opened to me; and I shall take these sources in the order in which they come before me; the evidence is therefore necessarily of a varied character, and it will I think be found conclusive. Besides those sections which are immediately devoted to the subject of structure, the paper contains others on the cause of the *flattening* of the air-bubbles in glacier ice, on the problem of glacier motion, and on the origin and cause of the Dirt-bands of the Mer de Glace.

§ 2. *General Aspect of the Veined Structure.*

The general appearance of the veined structure is well known. The ice of glaciers, especially midway between their mountain sources and their extremities, is of a whitish hue, owing to the number of small air-bubbles enclosed in the mass—the residue, doubtless, of that air which was originally entangled in the snow of which the ice is composed. Through the general whitish mass, however, at some places, blue veins are drawn, so numerous indeed, in some cases, as to cause the blue ice to predominate over the white. A laminated appearance is thus conferred upon the ice, the cause of the blueness being, that for some reason or other, the bubbles distributed throughout the general mass do not exist in the veins, or exist there in much smaller numbers.

In different glaciers, and in different portions of the same glacier, these veins exist in different stages of perfection. On the clean walls of some crevasses, and in the channels worked in the ice by glacial streams, they present a most beautiful appearance. They are not to be regarded as a partial phenomenon, or as affecting the constitution of glaciers to a small extent only. Vast masses of some glaciers are thus affected: by far the greater part of the Mer de Glace, and its tributaries, is composed of this laminated ice. The lower portion of the glacier of the Rhone, from the base of the ice cascade downwards, is entirely composed of it, and numerous similar cases might be cited.

To observe the structure of a glacier it is not even necessary to see the blue veins. Those who have ascended Snowdon, or wandered among the hills of Cumberland, or even walked in the environs of Leeds or other towns in Yorkshire and Lancashire, where the stratified sandstone of the district is used for architectural purposes, will have observed the exposed edges of the slate rocks, and of the stratified sandstone, to be grooved and furrowed by the action of the weather. In fact some portions of such rocks withstand the action of the atmosphere better than others, and these more resisting portions stand out in ridges while the softer portions between them are worn away. An effect exactly similar is observed upon the surface of the glacier. The laminated ice, exposed to the sun, and to wasting atmospheric influences, melts in a manner similar to the wasting of the rocks; little grooves and little ridges are formed upon the surface of the glacier, the latter being due to the more resisting ice, while the grooves are pro-duced by the melting of the less resisting mass between them.

The consequence of this is, that the light dirt scattered by winds and avalanches over the surface of the glacier is gradually washed into the little grooves, thus forming fine lines, which to the practised eye are an infallible indication of the structure of the ice underneath. Visitors to the Jardin have ample occasion to observe these striæ, for they are finely shown upon the surface of the Mer de Glace between the Augle and Trela-porte. When they are followed until they are intersected by a fissure or a stream, it is seen that the superficial groovings always mark the direction of the veined structure within the glacier.

§ 3. *Structure and Stratification*:—*Marginal Structure.*

Opinions at present are very diverse as to the origin of these veins. Professor FORBES first regarded them as being caused by the freezing of water which filled fissures in the ice, but he now discards the notion of freezing, and supposes the " incipient fissures" to be closed by " time and cohesion." M. AGASSIZ despairs of rendering an account of them, but calls them " bands of infiltration." The Messrs. SCHLAGINTWEIT have also treated the question, but with no greater success. In the paper published by Mr. HUXLEY and myself, pressure is referred to as the probable cause of the phenomenon, but we were unable at the time to furnish proofs of this. Apart from those who give public expression to their views upon the subject, I know that there are many who reject the pressure theory, and adopt instead of it the explanation that the blue veins of the glacier are merely the continuation of the strata of the névé; a view which has recently been upheld by Mr. JOHN BALL in the Philosophical Magazine. The matter indeed so stands, that in a recent *résumé* of glacier investigations, Professor MOUSSON of Zürich omitted the subject of structure altogether, for the express reason that the question is still in complete obscurity.

I will not take up the time of the Society in discussing the vague, involved, and often absurd explanations which have been given of the blue veins of glaciers, but state broadly that the question now rests between the pressure theory, and that of stratifica-tion. Taking the parallel geological phenomena, the question then is, Does the veined structure of glaciers correspond to the *stratification* or the *cleavage* of rocks?

In reply to this question, I will remark, in the first place, that the veins are not always, nor even generally, such as we should expect from stratification. The latter ought to furnish us with distinct planes extending parallel to each other for great distances through the glacier; this is by no means the general character of the veins. We observe blue streaks, some a few inches, some a foot, and some several feet in length upon the walls of the same crevasse, and varying from a fraction of an inch to several inches in thickness. In many cases the blue spaces are definitely bounded, giving rise to the lenticular structure described by Mr. Huxley and myself, but more usually they lose themselves as pale washy streaks in the general mass of the white ice. In fig. 1, I have endeavoured to give an idea of a very common aspect of the veined structure. Such a structure is not that which we should expect from bedding.

Fig. 1.

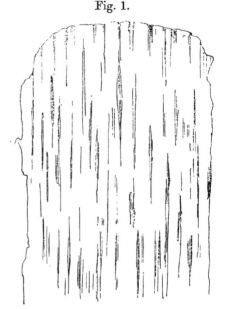

Again, taking the Glacier du Géant as a representative case, we have first of all the slopes of the Col du Géant, the collectors of the snow by which the glacier is formed. The fissures on these slopes exhibit beautifully, to a certain depth, the horizontal stratification. The lines of bedding may be seen as far down as the summit of the great ice-fall between the Rognon and the Aiguille Noire; and on the castellated masses at the summit of this fall, to which the name *seracs* has been applied, the lines of stratification may be distinctly seen. Escaped from the confusion of the fall, the glacier flows gently through a long valley towards its junction with the Léchaud and Talèfre at the Tacul.

Throughout the entire length of this glacier the planes of the structure are *vertical* or nearly so; sometimes they dip a little forward, but at other places they dip an equal quantity *backward*. Now let the mind figure, if it can, an agency which, as the mass descends the fall, shall turn up the horizontal strata of the Col du Géant and set them vertical, without a single break, throughout *the entire length* of the Glacier du Géant, and I imagine the effort to conceive of such an agency will be followed by the conviction that the change indicated is inconceivable.

Further, we often find, in the central portions of a glacier, the structure feeble, or scarcely developed at all, while at the sides it is well developed. This is often the case where the glacier moves through a valley of tolerably uniform inclination, and where no medial moraines occur to complicate the phenomenon. But if the veins mark the bedding, there seems to be no reason why we should not find them as clearly defined at the centre as at the sides; the fact, however, certainly is that we do *not* so find them.

Let me here show the true significance of this fact. If a plastic substance, such as mud, flow down a sloping canal, the central portions will flow more quickly than the lateral ones which are held back by friction. Now the flow may be so regulated that a circle stamped upon the central portion of such a mud-stream shall move downwards *without sensible distortion*, thus proving that the central mud is neither compressed nor stretched longitudinally; for if the former, the circle would be *squeezed* to an ellipse with its major axis transverse to the axis of the stream; and if the latter, it would be *drawn out* to an ellipse with its major axis parallel to the line of flow. A similar absence of longitudinal compression exists in many glaciers, and *in such ice-streams there is no transverse central structure developed.*

But let a circle be stamped upon the mud-stream near to its side; owing to the speedier flow of the centre, this circle must be distorted to an ellipse, because the part of the circle furthest from the side moves more quickly than the part nearest the side. Hence we shall have an ellipse formed with its major axis inclined downwards, indicating that the mud is compressed in one direction and expanded in another. An exactly similar state of things occurs in many glaciers; the ice near the sides is subjected to a pressure and tension like that here indicated, and we have marginal crevasses as the result of the tension, while the veined structure is, at all events, found associated with the pressure.

Fig. 2 will perhaps render my meaning more intelligible, in which cb, cb represents the sides of a glacier moving in the direction of the arrow. Here, while the central circle retains its shape, the side ones are squeezed and drawn out to ellipses. Marginal crevasses occur parallel to the lines mn, or perpendicular to the tension, while the dotted

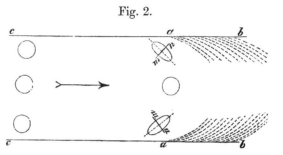

Fig. 2.

lines mark the direction of the blue veins which are at right angles, or nearly so, to the crevasses. I have dotted the line marking the direction of the structure along the margins ab, ab. In connexion with this point, I would refer to the instructive papers of Mr. HOPKINS[*], who has shown that in glaciers which move through valleys of uniform width, the directions of maximum pressure and tension are at right angles to each other, each of them enclosing an angle of 45 degrees with the side of the glacier.

I have simply said that the structure in the case described is "associated with the pressure;" thus confining myself within the strict limits of the facts. But what has been said shows that the pressure theory affords, at all events, a possible solution of a difficulty, which, without violence to fact, is inexplicable upon the hypothesis of stratification; the difficulty, namely, that a finely developed structure often exists along the margin of a glacier, while it is excessively feeble, or entirely absent, in the central portions.

§ 4. *Transverse Structure.—Glaciers of Grindelwald, the Rhone, &c.*

In many cases, however, the structure is not thus limited to the margins, but sweeps across the glacier from side to side, without interruption, being as well developed at the centre as at the margins. The stratification theory is wholly incompetent to account for this; the pressure theory requires that to produce this transverse structure the glacier must, at some portion of its route, have been forcibly compressed longitudinally. It was not till after my return from the Mer de Glace in 1857, that the full mechanical significance of *a change of inclination* in the glacier occurred to me.

Bend a prism of glass, we have compression on one side and extension on the other,

[*] Philosophical Magazine, 1845, xxvi. p. 148. See also Proceedings of the Royal Institution, vol. ii. p. 324.

with a neutral axis between, the mechanical conditions of the mass being shown by its action on polarized light.   The same is true of any other substance,—the concave surface of the bent prism is compressed.   Now at the bases of steep glacier slopes, where the inclination suddenly changes, we have a case of this bending, and along with it a thrust of the mass behind.   The concave surface is turned towards us, and that surface is thrown into a state of compression corresponding to the thrust, and to the change of inclination.   Hence it occurred to me, that the bases of the ice-falls, where the requisite change of inclination occurs, were likely to be the manufactories of the transverse structure.   The experience of 1858 completely verified this idea.

In illustration of my position I will take a representative case; and to render my observations capable of being easily checked, I will choose one of the most accessible glaciers in the Alps,—the lower glacier of Grindelwald.   One portion of this glacier descends from the Viescherhörner; but there is another branch which descends from the Schreckhorn, Finsteraarhorn and Strahleck, and it is to this latter branch that I now wish to direct attention.

Walking up this glacier from its place of junction with the tributary from the Viescherhörner, we come at length to the base of an ice-fall which forbids further advance upon the ice.   Let the glacier be here forsaken, and let the flanking mountain side, either right or left, be ascended, until a position is attained which affords a complete view of the fall and of the glacier stretching downwards from the base of the fall. The view from such a position will furnish a key to the development of the transverse structure.

It is, in point of fact, a grand *experiment* which Nature here submits to our inspection.   The glacier, descending from its névé, reaches the summit of the fall and is broken transversely as it crosses the brow.   It descends the fall as a succession of broken cliffy ridges, with transverse hollows between them.   In these latter the ice débris and the dirt collect, partially choking up the fissures formed in the first instance. Carrying the eye downwards along the fall, we see, as we approach the base, these sharp ridges toned down, and a little below the base they dwindle into rounded protuberances which sweep, in curves, across the glacier.   At the centre of the fall there is not a trace of the true structure to be observed.   At the base of the fall it *begins* to appear,—at first feebly, but soon becomes more pronounced; until finally, at a short distance below the fall, the eye can follow the structural groovings right across the surface of the glacier, while the mass underneath has become correspondingly laminated in the most beautiful manner.

It is difficult to convey, by writing, the force of the evidence which the actual observation of this great experiment places before the mind.   The ice at the base of the fall has to bear the powerful thrust of the descending mass; but more than this, the sudden change of inclination which it suffers throws its upper portion into a state of violent longitudinal compression.   The protuberances are squeezed more closely together, the hollows between them wrinkle up in submission to the pressure; the whole aspect of the glacier here gives evidence of the powerful exertion of the latter force; and exactly at the place where *it* is exerted, the structure makes its appearance, and being once manufactured, is sent onwards, giving a character to other portions of the ice-stream which have no share in its production.

An illustration, perhaps equally good and equally accessible, is furnished by the glacier of the Rhone. Above the great icefall which the traveller descending from the Furca has to his right, the horizontal bedding is exhibited in a more or less perfect manner, to a certain depth, upon the walls of the huge and numerous crevasses here existing. I have also examined this fall from both sides, and an ordinary mountaineer will find no difficulty in reaching a spot nearly opposite the centre of the fall, from which both the fall itself and the glacier below it are distinctly visible. Here a similar state of things to that already presented to his view reveals itself. The fall is *structureless*; the cliffy ridges are separated from each other by transverse hollows, following each other in succession down the slope; those ridges are toned down to protuberances at the base of the fall, becoming more and more subdued, until low down the glacier the transverse swellings disappear. As in the case of the Grindelwald glacier, the squeezing of the protuberances and of the spaces between them is visibly manifested. *Where this squeezing commences the transverse structure also commences*, and in a very short distance reaches perfection. All the ice that forms the lower portion of the glacier has to pass through this *structure mill* at the base of the fall, and the consequence is that *it is all laminated*.

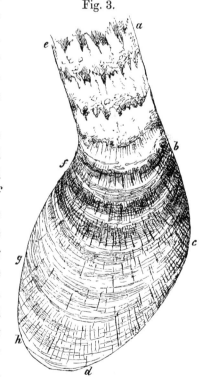

Fig. 3.

The case will be better appreciated by reference to figs. 3 and 4, the former being a sketch, in plan, and the latter a sketch in section of a part of the ice-fall and of the lower portion of the glacier of the Rhone. *a e b f* is the gorge of the fall, and *f b* its base. The transverse ridges are shown crossing the fall, being subdued at the base to protuberances, which gradually disappear further down the glacier. The "structure" sweeps across the glacier in the direction of the fine curved lines. On the plan I have also endeavoured to show the radial crevasses of the glacier; they are at right angles, or nearly so, to the structure. As would be inferred by those acquainted with what I have already written upon the influence of curvature, the side *b c d* of the glacier is much more violently crevassed than the side *f g h*.

Fig. 4 shows the cliffy ridges of the fall, and of the rounded protuberances below it, in section. The shading lines below denote the structure. The protuberances are so powerfully squeezed in some cases that they scale off at their surfaces.   Fig. 5 is a representation of such scaling off which I have observed at the bases of several cascades,

Fig. 4.

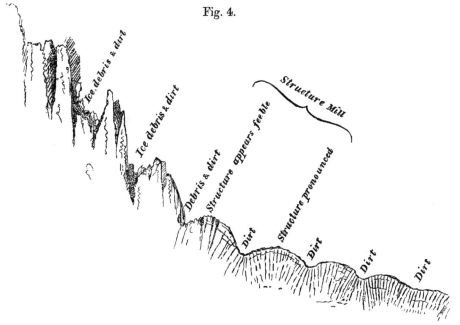

including those of Grindelwald, the Rhone, the Rognon, and the Talèfre, each of which has also its " structure mill" at the base of its cascade.   Fig. 5a is an example of scaling off which I have produced by artificial pressure.

Fig. 5.   Fig. 5 a.

It is to be borne in mind that the structure once formed prolongs itself into places which have no part in its formation; it would therefore be hasty to infer the relationship of structure and pressure from an observation of them at a particular portion of the glacier.   I have sometimes seen the veined structure parallel to the crevasses for a short distance: there are some transverse crevasses on the Glacier du Géant a little above Trelaporte which illustrate this; but it would be altogether erroneous to infer from this that the law which makes the structure perpendicular to the pressure, and hence, as a general rule, transverse to the crevasses, finds an exception here.   It is perfectly manifest that the structure which is brought into this unusual relationship to the crevasses has been developed far higher up; that the change of conditions from longitudinal pressure to longitudinal strain is too weak and transitory to obliterate it.   To effect obliteration, a force commensurate with that which produced the structure must be brought into play, and at the place now referred to no such force exists.

§ 5. *The Aletsch Glacier.*

Having made the foregoing observations upon the glacier of the Rhone, I proceeded to the Aletsch glacier, and during a residence of eight days at the hotel upon the slope of the Æggischhorn, made frequent excursions upon the ice. I had never previously seen this grand ice-stream, and my interest in it, at the time of my visit, was greatly augmented by the arguments which Mr. JOHN BALL had founded upon its deportment against the pressure theory of the veined structure. I shall here limit myself to a few brief remarks upon this subject.

I have already stated, and this must be particularly remembered, that the veined structure often appears in places which have no share in its production. The longitudinal structure in the centre of the stream of the Aletsch, for four miles above the base of the Æggischhorn, is not due to the lateral pressure endured by the glacier during these four miles. It is due, as Mr. BALL himself suggests, to the mutual thrust of the branch glaciers, which unite to form the trunk stream; and, once formed by this thrust, it perpetuates itself throughout a great portion of the trunk stream.

But it is urged against this view, that pressure exerted in new directions—the longitudinal pressure, for example, endured by the stream in its descent, and acting through long periods, ought,—if pressure has the power ascribed to it, to obliterate the first structure. Now here, again, it must be remembered that it is the portions of the ice near the bed of the glacier that yield, and that the upper portions of the ice, in many cases, are simply *floated* upon the moving under portions. Were the uniform "long reach" referred to by Mr. BALL strictly examined, it would, in all probability, be found that the ice near the surface is no more compressed than a log of timber would be if placed upon the glacier, and permitted to share its motion downwards.

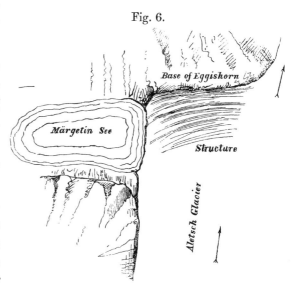

Fig. 6.

I may sum up by saying that a close examination of the glacier satisfied me, not only that it presented no phenomena which were at variance with the pressure theory, but also exhibited some which, as far as I could see, were perfectly fatal to the theory of stratification. The state of the ice at the base of the Æggischhorn, as shown in fig. 6, is certainly quite in harmony with the pressure theory; another fact observed upon the glacier shall be referred to at a future page.

§ 6. *Glaciers of Monte Rosa.*

I will next endeavour to describe the phenomena of structure exhibited in the system of glaciers in the neighbourhood of Monte Rosa. The general mechanical conditions of these glaciers will be evident to an observer stationed upon the Görner Grat, a point of view well known to travellers, and famous for the magnificence of the panorama which it commands.

As the observer stands here, facing Monte Rosa, the great Görner glacier, coming down from the heights of the old Weissthor at his left, flows beneath him. It is joined, in its course, by a series of glaciers from the sides of the opposite range of mountains. First of all comes the western glacier of Monte Rosa, which really ought to give its name to the trunk stream, as it is the most considerable of its tributaries. Into the glacier of Monte Rosa, and before the latter reaches the trunk valley, a glacier from the Twins, Castor and Pollux, pours its contents. Afterwards we have the Schwarze glacier, which lies between the Twins and the Breithorn; then the Trifti glacier, which lies upon the flank of the Breithorn, and afterwards the glaciers of the little Mont Cervin and of St. Theodule. The accompanying sketch (fig. 7) will render intelligible what I have to say regarding these glaciers.

Fig. 7.

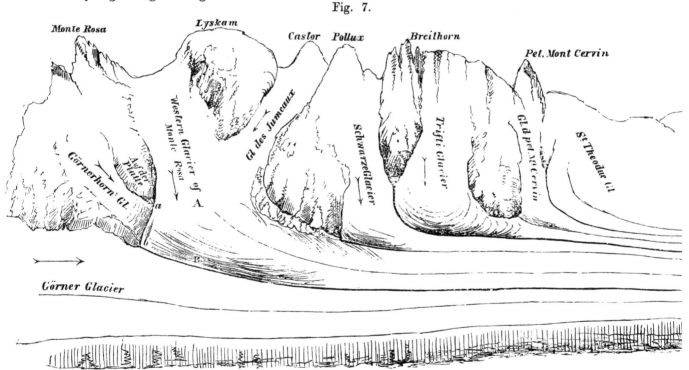

The small Görnerhorn glacier, which comes down the sides of Monte Rosa, is a very singular one. In comparison with the western glacier of Monte Rosa its mass is insignificant, and it is abruptly cut off by the latter along the line *a b*, a moraine occurring here, which may be regarded as forming, at once, the *terminal* moraine of the one glacier and the *lateral* moraine of the other. Thus the smaller glacier coming down the mountain side abuts against its more powerful neighbour, and we should infer from the inspection of the glacier that its terminus is subjected to great pressure.

Let the observer now suppose himself transported to the Görnerhorn glacier, at some distance above the terminal moraine *a b*; he will find there the transverse structure, if at all developed, excessively feeble and defective ; let him now walk downwards towards the moraine *a b*: every step he takes brings him to a place where the ice is subjected to a greater pressure, and every step also brings him to a better structure: both phenomena go hand in hand.   At the end of the glacier, alongside the terminal moraine and under it, the structure is finely developed.   If the observer now cross the glacier and ascend the rocks called *Auf der Platte*, from which he can command a near view of the Görnerhorn glacier, and embrace a large portion of it, he will be able to observe the gradual perfecting of the structure as the region of pressure is approached.   Towards the extremity of the glacier the surface becomes wrinkled, the groovings denoting the structure become more and more pronounced, the dirt striæ being more closely squeezed together ; and from these external aspects he may infer, with certainty, the gradual perfecting of structure within the glacier.

The western glacier of Monte Rosa next commands our attention.   This great stream occupies the valley between Monte Rosa and the Lyskamm, receiving the snows of the opposite sides of both.   The branch of the Görner glacier coming down from the Weissthor throws itself across the flow of its powerful neighbour, and deflects the latter, both of them afterwards moving together down the trunk valley, with a moraine, as usual. between them.

Before quitting the " Platte," we will suppose that the observer has endeavoured to form some idea of the mechanical conditions of the Monte Rosa glacier.   He would see the mass arrested in its descent by the Görner glacier, and compelled to accompany the latter.   A certain component of the weight of the glacier is borne by the ice where it comes into contact with the Görner glacier.   The observer would infer, from mere inspection, that if the structure be due to pressure, it ought to be most fully developed near the moraine which separates the Monte Rosa from the Görner glacier.

If he now pass from the " Platte " to the ice, and cross to the centre of the Monte Rosa glacier to A, he will find the structure there excessively feeble, if at all developed. Let him now walk straight down the glacier towards B, where the pressure is most intense.   Every step he takes downwards brings him to more perfectly veined ice ; and I am not acquainted with a more splendid example of laminated structure than that exhibited by this glacier along the moraine, and for some distance from it, at its southern side.

The system of glaciers which next come under review are exceedingly instructive. In no place in the whole range of the Alps are the effects of pressure and the phenomena of structure more strikingly exhibited.   I have endeavoured, in the sketch, to render the aspect of these glaciers intelligible.   The Schwarze glacier moves down a steep mountain slope, and welds itself to the Monte Rosa glacier at the bottom.   But the great mass of this latter enables it to pursue its way without being compelled to swerve sensibly by its feebler neighbour.   The latter is forced to bend abruptly. and from a wide irregularly-shaped field of névé, it is squeezed between the Trifti and the Monte Rosa glaciers to the narrow band represented in the figure, and moves thus downwards.

The Trifti glacier itself is perhaps a still more striking illustration of the power of ice to yield when subjected to pressure for a long period. The aspect of the real glacier is much the same as that shown upon the sketch. *It* also is compelled to change its direction, and to flow as a narrow stripe along the trunk valley, being hemmed in between the strip of the Schwarze glacier and that of the glacier of the little Mont Cervin. A beautiful system of bands is to be seen at the lower portion of this glacier.

The inspection of the sketch will show better than words the modifications of shape which the lower portion of the glaciers undergo by the pressure of their higher portions, and the resistance of the trunk stream. They are turned aside, firmly welded together, and form a series of parallel narrow bands, separated from each other by moraines. They are all well seen either from the Görnergrat or the summit of the Riffelhorn.

I have examined each of these glaciers, and find the same to be true of all of them. High up the structure is feeble; as we descend it becomes more pronounced, and at the places where the tributaries join the trunk, and the ice has to bear the full thrust of the mass behind it, we have a finely developed structure.

§ 7. *Coexistence of Structure and Stratification.—The Furgge Glacier.*

The evidence of the association of pressure and glacier lamination which I have thus far laid before the Society, will, I think, be admitted to be very strong. I have no hesitation in saying that the stratification theory has nothing to urge at all to be compared with it in point of cogency. Still I cannot help feeling how a critical and well-informed mind might weaken the force of what I have adduced. Difficult as the conception is, it might be urged that the structure, so fully developed near the margins of glaciers, may be due to a turning up of the strata edgeways, in consequence of a wide névé being squeezed into a narrow channel,—just as a sheet of paper, if forced through a groove less than itself in width, would turn up at its edges. It might also be urged that the structure developed alongside and under the medial moraines, is due to the placing side by side of these folded-up strata; the perfect welding of both and the clearer development of the structure being conceded as possible consequences of the mutual pressure. This indeed is Mr. BALL's view of the subject; and M. AGASSIZ assumes such a folding-up of the strata of the Unteraar glacier. With regard to the transverse structure also, it might be said that we do not know how the interior of the mass is affected in descending the ice-falls. The mind, it is true, finds great difficulty in conceiving of any agency which could set the strata which were horizontal above the fall, vertical below it; still this difficulty may be due to our ignorance of the mechanical conditions of the mass during its descent. In this way it would be quite possible to fritter away conviction to a mere opinion, and hence arose a strong desire on my part, either to confirm these surmises, or to place the pressure theory, once for all, beyond the power of such attacks.

One conclusive observation is still wanting to establish the analogy between glacier lamination and the cleavage of slate rocks. In the latter case the arrangement of the strata has been traced by their organic remains; and, indeed, stratification has often been visibly exhibited coexistent with cleavage, both crossing each other at a high angle. If a similar state of things could be detected upon a glacier, it would at once lay the axe to the root of all the scruples above referred to, and place the pressure theory upon an unassailable basis. The consciousness of this was sufficient to stimulate me in the search of such evidence.

I had visited all the glaciers hitherto mentioned, and others not mentioned, without obtaining more than one clear case of the kind: this case I observed upon the Aletsch glacier on the 6th of August. Not far from the junction of the Middle Aletsch glacier with the trunk stream, a crevasse exposed a wall of ice 50 or 60 feet in height, upon which the stratification was exposed, and cutting the stratification at a high angle were the groovings which marked the true veined structure. The association was distinct; my friend Professor RAMSAY was with me at the time; I drew his attention to the fact, and to him the case was perfectly conclusive. Thus the Aletsch glacier, which had been referred to by Mr. BALL as furnishing evidence against the pressure theory, gave us a fact, which, as far as I could see, was perfectly fatal to the theory of stratification.

But the case was solitary, and although inspiriting at the moment, its effect upon the mind became feeble as time passed, and no repetition of the observation occurred. I had remained at the Riffel from the 9th to the 18th of August, exploring all the adjacent glaciers, and adding each day to my stock of knowledge; but I met no case in which the structure and the bedding were so clearly and independently exhibited, as to leave an adherent of the stratification theory no room for doubt. Wednesday the 18th of August was to be my last day at the Riffel, and it was devoted to the examination of the Furgge glacier, which occupies the space between the pass of St. Theodule and the Matterhorn.

Crossing the valley of the Görner glacier, I climbed the opposite mountain slope, and passing the Schwarze See, soon came upon the glacier referred to. I walked up it until I found myself in a kind of *cul de sac*, flanked by precipitous ice-slopes, and opposed in front by a cascade composed of four high terraces of ice. The highest terrace was composed principally of broken cliffs and peaks of ice, and it had let some of its frozen boulders fall upon the platform of the second terrace, where they stood like rocking-stones on the point of falling. The whole space at the foot of the fall was covered with quantities of crushed ice, while some coherent masses, upwards of 200 cubic feet in volume, were cast to a considerable distance down the glacier.

Upon the face of the terraces the stratification of the névé was beautifully shown. Above the fall the névé extends as a frozen plain, quite undisturbed, so that the bedding took place with great regularity; and being broken through for the first time at the summit of the fall, the lines of stratification were peculiarly well defined and beautiful.

Towards the right of the fall, looking upwards, this was particularly the case; for here no pressure had been exerted upon the beds sufficient to contort them or to rupture their continuity.

The figure of a vast lake pouring its waters over a rocky barrier, which curves convexly upwards, thus causing the water to rush down it, not only longitudinally over the vertex of the curve, but also laterally over its two arms, will convey to the mind a tolerably correct conception of the appearance of the fall. Towards the centre the ice was powerfully squeezed; the beds were *bent*, and their continuity often ruptured, so as to exhibit faults; but they were as plain, and as easily traced, as in any other portion of the fall. I thought I saw structural groovings running at a high angle to the stratification. Had the question been an undisputed one I should have felt *sure* of this, for the groovings were such as always mark the structure. The place being dangerous, I first observed it from a little distance through my opera-glass; but at length, resigning the instrument to my guide, and leaving him to watch the tottering blocks overhead, and to give me warning in case of their giving way, I went forward to the base of the fall, peeled the grooved surface away with my axe, and *found the true veined structure underneath*, running, in this case, nearly at right angles to the stratification.

The superficial groovings were not uniformly distributed over the whole face of the terrace, but occurred here and there where the ice had yielded most to the pressure. I examined several of these places, and in each instance found the superficial grooving to be the exponent of the true veined structure underneath, the structure being in general nearly *vertical*, while the lines of bedding were *horizontal*. The coarse bands which marked the division of the beds were also seen underneath, when the surface of the ice was removed. Having perfectly, and with deliberation, satisfied myself of these facts, I made a speedy retreat; for the ice blocks were most threatening, and the time of day that at which they fall most frequently.

We now resolved to try the ascent of the glacier to the right; it was much riven, but perfectly practicable to a good iceman. To me it was also perfectly delightful; in fact, as regards the relationship of structure and stratification, this glacier taught me more than all the others I had visited taken together. Our way lay through fissures which exposed magnificent sections, and every step forward added further demonstration to what I had already observed at the base of the fall. The bedding was perfectly distinct, and the structure equally so, the one being at a high angle—sometimes at a right angle—to the other. Among these crevasses the pressure was in some cases greater than on the fall, and the structure proportionally more pronounced. The crumpling of the beds demonstrated the exercise of the pressure, and the structure went straight through such crumplings, thus furnishing me with numerous parallels to the case observed by Professor SEDGWICK, Mr. SORBY, and others, of the passage of slaty cleavage through contorted beds. Indeed I question whether the phenomena of cleavage and bedding, in the case of slate rocks, were ever exhibited, side by side, with a distinctness equal to that of the stratification and "structure" of ice in the present instance.

Fig. 9 represents a crumpled portion of the ice, with the lines of lamination passing through those of bedding at a high angle.   Fig. 10 represents a case where a fault

Fig. 10.

Fig. 9.

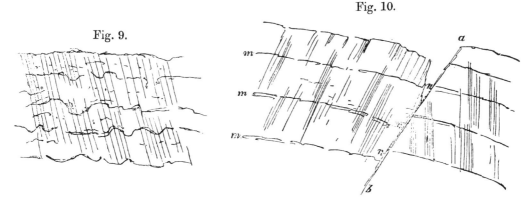

occurred, the veins at both sides of the line of dislocation *ab* being inclined towards each other.   The lines *mn*, *mn* represent of course the lines of bedding, and the lines crossing them the structure.   These observations are conclusive as regard the claims of the rival theories of structure and stratification*.

§ 8. *On the White Ice-seams of the Glacier du Géant, and their relation to the Veined Structure.*

From an elevated point at Trelaporte I observed a remarkable system of white bands sweeping across the Glacier du Géant in the direction of the structure.   From one of the moraines near the junction of the three tributary glaciers, the same system of bands present a very striking appearance.   They consist of a hard white ice, more resistant than the general mass of the glacier, and in some cases rising to a height of three or four feet above the surface.   On close examination I found that they penetrated the

Fig. 11.

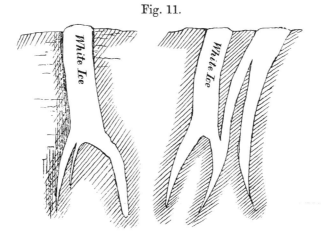

glacier only to a limited depth.   In fig. 11 I have given the sections of two of these

---

* While correcting this proof, I find a case of this kind figured by M. Agassiz in the atlas to his 'Système Glaciaire,' pl. 8. fig. 3.

veins, about 15 feet deep, which were exposed on the walls of a crevasse high up the glacier. They constituted a kind of *inverted glacier trap*, and I was led to a knowledge of their origin in the following way.

In one of my earliest visits to the base of the ice-fall of the Talèfre, I observed a curious disposition of the veined structure on the walls of some of the crevasses: fig. 12 represents one case of the kind, and fig. 13 another, and numerous similar ones find a place

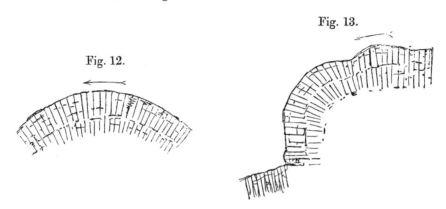

Fig. 13.

Fig. 12.

in my note-book. In the former case the veins fell *backward* as well as forward, being vertical through the central portion of the curve. In fig. 13 the position of the veins varies in a very short distance from the vertical to the horizontal.

I found that the portions of ice which showed the phenomena, formed, when seen from a point of view sufficiently commanding, a part of a system of crumples or protuberances which swept round the base of the fall, between the moraine which descends along it from the Jardin, and its highest lateral moraine. I have already referred to the protuberances which sweep across the Strahleck branch of the Lower Grindelwald glacier, and of those of the glacier of the Rhone: those to which I now refer were of the same character.

Right and left from the position where the crumples were most pronounced they gradually became subdued, shading off to a mere undulating surface; the sides of a crevasse intersecting this surface longitudinally presented the structural arrangement shown in fig. 14. It will be observed that the directions of the veins change in accordance with the undulations of the surface.

Supposing the squeezing of the mass to become so violent that the gentle undulations shall become steep crumples, the deviation of the structure from parallelism with itself would, of course, be augmented. This prepares us to understand the exact phenomena observed at the base of the Talèfre cascade. Fig. 15 represents a series of crumples following each other in succession at the place referred to; at the base of each I found a vein of *white* ice, *a, a*, wedged into the mass. This interrupted the continuity of the structure; the abrupt change in its direction at opposite sides of the white band being. as shown in the figure, in every case observed.

I found that the width of the seams was exceedingly irregular, varying, at different

portions of the same seam, between 6 inches and 3 or 4 feet.   I also found that a seam
sometimes became forked so as to form two branches, which thinned gradually off until

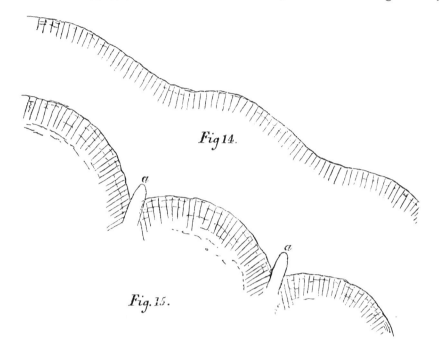

Fig 14.

Fig. 15.

they finally vanished.   Fig. 16 is an example of this kind: the seam was divided
at the point *a*; one of its branches *ran up
the face of the crumple*, thinned off and disappeared at *b*; the other widened considerably, but finally thinned off and also
vanished.

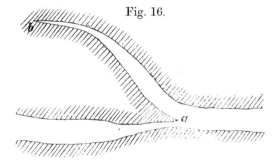

Fig. 16.

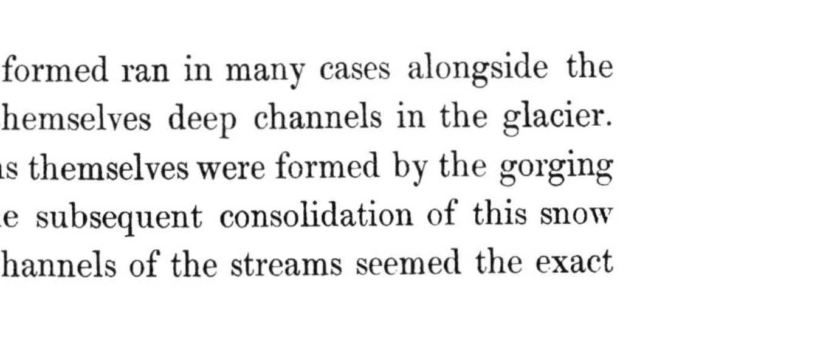

Along the bases of the crumples the fil-
lets of water which poured down their faces
were collected and flowed.   The streams thus formed ran in many cases alongside the
existing veins of white ice, and had worn for themselves deep channels in the glacier.
The thought soon suggested itself, that the seams themselves were formed by the gorging
up of those channels by snow in winter, and the subsequent consolidation of this snow
during the descent of the glacier.   Indeed the channels of the streams seemed the exact
matrices of the seams of white ice *.

---

* The fact of one branch of a vein running up the face of a crumple, seems to prove that the ice, which
at one time constitutes the *base* of a crumple, does not always remain so; the bases of the crumples are
sometimes *lifted up* by the squeezing.   The *horizontal* structure at the fronts of many of the crumples
seems due to a local forcing forward of one protuberance over that next below it.   Were the matter tested
by strict measurement, I think it would be found that different portions of the crumples move downwards
with different velocities.   According to this view, upon the general motion of the glacier there are *local*
motions superposed.

I afterwards traced the seams of white ice of the Glacier du Géant to their origin amid the ridges and hollows at the base of the great ice-fall of Le Rognon. In some cases the seams opened out into two branches, which, after remaining for some distance separate, would unite again so as to enclose a little glacial island; at other places lateral branches were thrown off from the principal seam, presenting the form of a glacier stream which had been fed by tributary branches. Fig. 17 is the plan of an actual

Fig. 17.

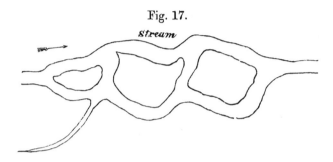

stream observed at the base of the ice-fall; fig. 18 is the plan of a seam of white ice

Fig. 18.

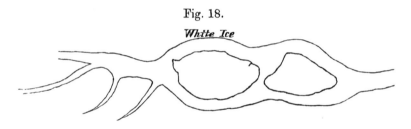

observed the same day lower down the glacier; their relationship is evident. I may remark that I have observed other seams produced by the gorging of *crevasses* with snow, and the subsequent closure of the fissures.

Considering the place where they are formed, these channels cannot escape compression; but let me remove all uncertainty on this point, by proving that not only at the base of the seracs, but throughout almost its entire length, the Glacier du Géant is in a state of longitudinal compression.

The first proof I have to offer is that the transverse undulations of the glacier, to which reference has been so often made, become gradually *shorter* as they descend. A series of three of them, measured along the axis of the glacier on the 6th of August 1857, gave the following respective lengths—955, 855, and 770 links, the shortest undulation being the furthest down the glacier. Now these undulations, as I shall subsequently show, are due to a regularly recurrent action, and are doubtless originally of the same length; that the lower ones are shorter than the higher must therefore be due to compression.

The following observation is, however, more conclusive. About three-quarters of a mile above the Tacul, and to the left as we ascend, there is a green patch upon the craggy mountain side. From this spot, as a station, I set out with a theodolite a line (No. 1) transverse to the axis of the glacier.

From a station lower down, chosen in a couloir along which the stones are discharged from the end of a secondary glacier which hangs upon the slope of Mont Tacul, I set out a second line (No. 2) transverse to the axis of the glacier.

A third line (No. 3) was set out across the glacier about a quarter of a mile still lower*.

The mean daily motion of the centres of these three lines is given in the annexed Table, and also their distances apart.

Mean daily motion of three points upon the axis of the Glacier du Géant.

	inches.	Distances apart.
No. 1	20·55	} . 2477 links.
No. 2	15·43	} . 2215 links.
No. 3	12·75	}

The advance of the hinder lines upon these in front is most strikingly shown by these measurements; and the proof that the Glacier du Géant is in a state of longitudinal compression is thus complete.

Here then we have a vast ice press, and here we have the pure snow filling the transverse channels of the streams. We are thus furnished with an experimental test on a grand scale of the pressure theory of the veined structure. In 1857 I examined a great number of these seams of white ice, and found in many of them *a finely developed lenticular structure*. In 1858 I also examined the seams, and found some of them " ribboned" in the most exquisite manner by the blue veins; indeed I had never seen the veins more sharply and beautifully developed.

This structure was observed in portions of the seams at and near the centre of the glacier, where the differential motion observed at the sides does not exist. This fact, I think, throws grave difficulties in the way of any theory which makes the veined structure *dependent on differential motion*, and more especially a theory which requires " *a very considerable amount of this differential motion* to produce any sensible degree of stratification in the vesicles."

§ 9. *On the flattening of Air-bubbles in Glacier Ice, and its relation to the Veined Structure.*

Those who have given their attention to the subject, know that the bubbles contained in glacier ice are, in general, not spherical, but *flattened*; and that from their shape conclusions of the greatest import have been drawn regarding the internal pressures of glaciers.

M. AGASSIZ draws attention to this subject in the following words:—" The air-bubbles undergo no less curious modifications; in the neighbourhood of the névé, where they are most numerous, those which one sees on the surface are all spherical or ovoid, but by degrees they begin to be flattened, and near the end of the glacier there are some that are so flat *that they might be taken for fissures when seen in profile*. The drawing.

* These three lines are drawn upon the sketch map at page 268 of Part I. of these researches (FF', GG', HH').

fig. 10, represents a piece of ice detached from the gallery of infiltration; all the bubbles are greatly flattened.   But what is most extraordinary is, that far from being uniform, *the flattening is different in each fragment*; so that the bubbles, according to the face which they offer, appear either very broad or very thin.   I know of no more significant fact than this, *since it demonstrates that each fragment of ice is capable of undergoing in the interior of the glacier a proper displacement independently of the movement of the whole.*

"The same flattening of the bubbles," continues M. AGASSIZ, "is found at a greater depth.   While engaged in my boring experiments, I observed attentively the fragments of ice brought up by the borer.   I found in them almost flat bubbles, perfectly similar to those of the fragment figured above, at all depths from 10 to 65 metres.   It follows *hence* that *a strong pressure is exercised on the interior of the glacier.*"

The description of the "flattening" here given is correct: all observers agree in corroborating it, and every observer with whom I am acquainted draws substantially the same conclusion from the phenomenon that M. AGASSIZ does.   Professor THOMSON's speculation upon the subject is particularly refined and ingenious.

Mr. JOHN BALL converts the flattening of the bubbles into evidence against the pressure theory of the structure in the following way:—"As AGASSIZ has pointed out," writes Mr. BALL, "and I have frequently verified his observations upon this point, though the air-cavities show traces of compression reducing them to the form of flattened lenses, the directions in which they are flattened are most various, *and show no constant relationship to the planes of the veined structure.*   Here then we have direct evidence that separate portions of the ice have been acted upon by pressure sufficient in amount to modify their internal arrangement, but that these pressures have not acted in the same, or nearly the same direction."

Granting the inference that the observed flattening "furnishes direct evidence" of pressure, the foregoing argument would, I confess, be a very formidable one.   If the bubbles are thus flattened by pressure, and if the veined structure, as I contend, be the result of pressure, and approximately at right angles to the direction of the force, we ought to have the bubbles squeezed out in planes parallel to the structure.   The fact that the bubbles are not so squeezed out, would then afford a strong presumption that the structure is not produced by pressure.   I expect, however, to be able to prove that the shape of the bubbles is *not* a "direct evidence" of pressure, as hitherto assumed ; and I think, as I do so, it will be seen how necessary it is to associate experiment with an inquiry of this kind, if we would read aright our observations.

In a paper in the Philosophical Transactions on the Physical Properties of Ice, I have shown that when a sunbeam traverses a mass of ice, the latter melts at innumerable points in the track of the beam, and that each portion melted assumes the form, not of a globule, but of a flower of six petals.   The planes in which these flowers are formed are independent of the shape of the mass and of the direction of the beam through it; they are always formed *parallel to the surface of freezing.*

This is a natural consequence of the manner in which the particles of ice are set together by the crystallizing force.   By the slow abstraction of heat from water its particles build themselves into these little stars, and by the introduction of heat into a mass so built the architecture is taken down in a reverse order.   In watching the formation of artificial ice, by the machine of Mr. HARRISON referred to in my paper, I have seen little solid stars formed, by freezing, which were the exact counterparts of the little liquid stars formed by melting.   So far as I can see, the complementary character of the phenomena is perfectly natural, and presents no difficulty to the mind in conceiving of it.

When the beam is intense, and its action continued for some time, the flowers expand, so as to form liquid plates within the mass.   Looked at edgeways, these liquid spaces appear like fine lines; which proves that the melting is not symmetrical laterally and vertically, but that the ice melts in the planes of freezing much more readily than at right angles to these planes.

If an air-bubble exists within ice, and if the ice melts at the concave surface of this bubble, as might be expected from the foregoing facts, the ice will so yield that the composite cell of air and water will not be spherical, even though the bubble of air may originally have been so.   In the planes of freezing the mass yields most readily, and the cavity containing the air and water will appear *as if flattened by a force acting perpendicular to these planes.*   This is not a deduction merely, but an observation which I have made in a hundred different cases.

What I have here said applies to ordinary lake ice; but glacier ice has no definite " planes of freezing."   The substance is first snow, which sometimes, it is true, falls regularly in six-rayed crystals, as observed by myself on the summit of Monte Rosa; but it is usually disturbed by winds, while falling, and whirled and tossed by the same agency after it has fallen; the mountain snow is often melted, mixed with water and refrozen.   Even after it has become consolidated it is often shattered in descending precipitous slopes.   In such ice definite planes of crystallization are, of course, not to be expected.

If we suppose a mass of lake ice to be broken up into fragments, and these fragments thrown together confusedly and regelated in their new positions to a continuous mass, we have an exact image of the character of the glacier ice in which this flattening of the bubbles in different directions has been observed.

In the paper already referred to, I have given a sketch of a piece of ice composed of such segments, and have described the effects obtained with it.   That ice was sold to me as Norway lake ice.   I am not aware whether glacier ice is ever imported into this country from Norway; but if it be, the piece in question must, I think, have belonged to it.   It is so like all the glacier ice that I have examined since that time, and so unlike all the lake ice, that I feel little hesitation in saying that it belonged to the former*.   No matter how coherent and optically continuous a mass of ice may be, a condensed sunbeam would at once tell us whether it belonged to a lake or to a glacier.

* Perhaps formed from the connecting together of confused fragments.

I have given in fig. 19 a sketch of a piece of ice taken from the end of the great Allalein glacier, on the Swiss side of the Monte Moro.

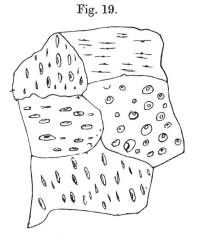

Fig. 19.

On reference to M. AGASSIZ's figure, it will be quite manifest that we are both dealing with the same phenomenon; we have the division of the ice into "angular fragments," the flattening of the "bubbles," and the non-parallelism of their directions in the different fragments.

Fig. 20 is a sketch of a piece of ice which showed the veined structure. The line AB was parallel to the veins, and it will be seen that the "bubbles" are inclined to this line at different angles, and in different azimuths. The circles indicate, of course, that the "bubbles" were there parallel to the horizontal face of the slab, while the *lines* indicate that they were perpendicular. In one case the bubbles are seen in plan, in the other case in section. The ellipses show the bubbles foreshortened where their planes are oblique to the surface of the slab.

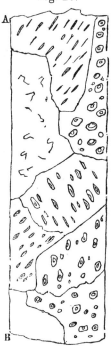

Fig. 20.

Associated with the air-bubbles, and usually beyond comparison more numerous in ice taken from the "ends" of glaciers, were the round liquid disks which I have described in my paper on the Physical Properties of Ice. Associated with each liquid disk was a *vacuous spot*, which shone with exceeding lustre when the sunbeams fell upon it. That the spots were vacuous, and not bubbles of air, I proved by permitting them to collapse under warm water; the collapse was complete, and no trace of air arose from them.

These, I doubt not, are the "bubbles" observed by M. AGASSIZ "near the end of the glacier," and which were "so flat that they might be taken for fissures when seen in profile."

These "vacuum disks," as I have usually called them, were invaluable as indicators of the planes of crystallization. When a condensed sunbeam was sent through the mass, the six-petalled flowers, which always indicate the planes referred to, started into existence parallel to the disks. Consequently, as the beam passed through different fragments, flowers were formed, in different planes, along the track of the beam.

True air-bubbles, associated with water, also occurred in these masses of ice, and such composite cells were always flattened out in the planes of the vacuum disks.

The fact then is that many of the so-called air-bubbles are not air-bubbles at all, and that the so-called "*flattening*" is in reality no flattening at all; and that pressure, in the sense hitherto conceived, has had nothing whatever to do with the shape of these bubbles. In glacier ice, as in lake ice, their shape is determined by the crystalline architecture. The conclusion that they were *squeezed* flat seems to have been drawn by M. AGASSIZ, and reproduced by subsequent writers, without due regard to the difficulties associated with it. That the pressures of a glacier are so parcelled out as to squeeze *contiguous fragments* of ice, not exceeding a cubic inch in size, in all possible directions, is so improbable, that reflection alone must throw great difficulties in the way of its acceptance.

It is with some diffidence that I here venture to express an opinion upon a question that I have not specially examined; but it appears to me probable that the decomposition of glacier ice into large granules, regarding which so much has been written, may be connected with the foregoing facts. The ice of glaciers is sometimes disintegrated to a great depth; causing it to resemble an aggregate of jointed polyhedra more than a coherent solid. I was very near losing my life in 1857 on the Col du Géant by trusting to such ice; and last summer I found vast masses of it at the end of the Allalein glacier. Blocks a cubic yard and upwards in volume, fell to pieces to their very centres on being overturned; they were an aggregate of granules, whose average volume scarcely exceeded a cubic inch. From the constitution which the foregoing observations assign to glacier ice, this disintegration seems natural. The substance is composed of fragments which are virtually crystallized in different planes; and it is not to be expected that the union along the surfaces, though they may be *invisible* when the ice is sound, is as intimate as that among the different parts of a mass homogeneously crystallized. Besides, ice no doubt, and all uniaxal crystals, expands by an augmentation of temperature, differently in different directions, and hence a differential motion of the particles on both sides of one of the above surfaces when the volume of the substance is changed by heat or cold is unavoidable. Such surfaces then would become surfaces of discontinuity, and perhaps produce that granular condition which has occupied so much of the attention of observers.

§ 10. *Physical Analysis of the Veined Structure.*

The relation of pressure and structure has been shown in the foregoing pages, but the mode in which the pressure acts remains yet to be considered. As regards their causes, slaty cleavage and slaty structure have been reduced to one and the same; but as regards the operation of that cause, no two things can, I imagine, in some respects at least, be more different.

In a note at page 336 of the 'Proceedings of the Royal Society' for January 1857, I refer to an experiment in which a clear mass of ice was caused by pressure to resemble a piece of fissured gypsum, and I there promised the full details of the experiment in due time. In my paper on the Physical Properties of Ice this promise is fulfilled; I have shown how a mass of compact ice may be liquefied by pressure, in parallel planes perpendicular to the direction of the force, and explained the effect by reference to the ingenious deductions of Mr. JAMES THOMSON from CARNOT's maxim.

Let the attention now be fixed on the state of a glacier at the base of one of the ice-falls where it is bent so as to throw its surface into a state of longitudinal compression. According to the above experiments, the glacier whose temperature is 32° FAHR. must here be liquefied *in flats*, perpendicular to its axis. A liquid connexion is thus established between all the air-bubbles which are intersected by any one flat, and a means for the escape of this air from the glacier is thus furnished. The water produced is also partially expressed, partially absorbed by capillary attraction of the adjacent bubble ice, and partially refrozen when the pressure is relaxed. It is, I think, perfectly manifest that such a process, each step of which may be illustrated by experiment, must result in the formation of the blue veins*.

All the experiments and observations recorded in the paper on the Physical Properties of Ice were made with reference to the glaciers, and one experiment there recorded illustrates the present point more forcibly than any words could do,—it is that in which I have thrown a prism of ice into a state of compression, which brings one side of it into the exact condition of the glacier at the base of an ice-fall. Fig. 21 is precisely the

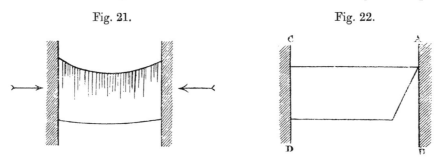

Fig. 21.   Fig. 22.

same as that given at page 226 of my paper, the prism being placed horizontal instead of vertical, so as to show its bearing upon the present point more distinctly. The original shape of the piece of ice is given in fig. 22, which by compression between the surfaces AB and CD is reduced to the shape and condition of fig. 21. The vertical lines represent the planes of liquefaction, and they correspond exactly to the planes of the blue veins in the glacier. The application of the same principles to all cases where pressure comes into play is sufficiently obvious.

In the experiments with the hydraulic press, the portion of ice between each two liquefied flats transmits the pressure without sensible yielding. We have no difficulty in conceiving that the same holds true of glacier ice, and that pressure may be transmitted through one portion of a bubbled mass of ice and produce the liquefaction of another portion, without sensible distortion of the bubbles contained in the former. One of the objections which have been urged against the pressure theory is thus, I conceive, completely answered,—the objection, namely, that the white ice which transmits the pressure ought to have its bubbles flattened. Indeed this objection continued to be of weight only so long as it was imagined that the observed flat bubbles had been *squeezed* to this shape; a notion, which I think will no longer be entertained.

---

* The mechanical actions which accompany the development of ordinary slaty cleavage, must, I think, also manifest themselves to some extent in the glacier.

§ 11. *Remarks on Glacier Motion.*

It is only by slow degrees that we master from actual observation, a problem so large as that presented by the glaciers; the muscular labour alone being such as to render the expenditure of a considerable amount of time unavoidable. The examination of the various questions connected with glaciers, has been therefore, in my case, distributed over some years, and not until last summer was I able to devote the requisite attention to the subject of the present section, which, however, is essential to a right comprehension of the physics of glacier motion.

It would be a problem eminently worthy of any geologist, to lay down upon a trust-worthy map of Switzerland the directions of the striæ on the rocks over which ancient glaciers have moved; and to one who sees its importance and desires exact information upon this subject, it must be a matter of surprise that nothing of the kind, in a systematic way, has yet been attempted.. A suitable map furnished with such lines of direction, carefully and conscientiously drawn, would impart more satisfactory information than all the volumes that ever have been, or ever will be written upon the subject. Here is a piece of work loudly calling for accomplishment, and one on which any young geologist may base an honourable reputation.

Mr. HOPKINS, I believe, was the first to urge the existence of *roches polies* at the ends of existing glaciers and along the continuations of existing glacier valleys as an evidence in support of the sliding theory. That such facts exist is known to every body, and that the rocks are thus polished and rounded by the glaciers sliding over them is incontrovertible. Let a traveller, if he wish to obtain a wealth of information upon this subject, transport himself to the terminus of the Unteraar glacier, and walk thence down the valley through which the river Aar now flows. On all sides he will obtain the most striking evidence that the base of the valley was once the bed of the glacier. The rocks are polished and striated, and present at some places the appearance of huge rounded mounds, which, at first sight, would appear to offer an insuperable barrier to the motion of the glacier, but which show by their aspect that the ice actually moved over them, grinding off their angles and furrowing their summits and sides. All along the valley towards Meyringen, similar evidences exist. In fact, the phenomenon is very common, and admitted on all hands.

The conclusion which Mr. HOPKINS has drawn from these facts is unavoidable; the glaciers must have *slidden* over the rocks on which such traces are left. To an eye a little practised in those matters, the precise limits reached by the ancient glaciers are perfectly visible. The junction of the rounded and abraded portions of the mountains, with those portions which in ancient times rose over the then existing ice, is perfectly distinct; and I should say in the valley of the Aar reaches to a height of more than a thousand feet above the present bed of the river. The valley of Saas, in the Canton de Valais, furnishes magnificent examples of the same kind.

At all places, from the base of the ancient glacier to its surface, sliding must have occurred; the evidence of it is perfectly irresistible. The summit of the Grimsel pass constituted the bed of an ancient névé; and the groovings and polishings, at the very summit of the pass, show that the ancient névés, as well as the ancient glaciers, slid upon their beds. In company with my friend Professor RAMSAY, and assisted by his great experience, I visited the sites of other ancient névés, and found the same true of all of them; they all slid more or less over their beds.

No investigator of glacier motion can shut his eyes to those facts, nor refuse to give them their proper weight. *The sliding theory is beyond doubt to some extent true*; and many of the objections raised against it, and still repeated in works intended to instruct the public, are altogether futile.

Here, as in other cases, we find that the extreme facts have been dwelt upon principally by rival theorists, and coexistent truths have, by partial treatment, been rendered apparently hostile to each other. It is perfectly certain that a glacier *changes its form* by pressure like a plastic mass, but it is equally true that it *slides over its bed*.

§ 12. *On the Dirt-bands of the Mer de Glace.*

In walking over the Mer de Glace, we soon observe differences in the distribution of the dirt upon its surface; but while standing on the glacier itself, no orderly arrangement of the dirty and clean spaces is observed. From a point, however, which commands a view of a large portion of the glacier, it is seen that the dirty spaces are arranged so as to form a series of broad brown curves, which follow each other in succession down the glacier. They were first observed by Professor FORBES from the heights of Charmoz, on the 24th of July, 1842, and from the same place, on the 16th of July, 1857, I observed them. Last summer I counted eighteen of them from the same position, and this agrees with the number observed by Professor FORBES. This agreement, after an interval of sixteen years, proves the regularity of their occurrence.

These bands were different from anything of the kind I had previously seen, and I felt that the explanation given of the "dirt-bands" observed by Mr. HUXLEY and myself would not completely account for those now before me. They were perfectly detached from each other, and resembled sharp hyperbolas with their vertices pointing downwards.

From Charmoz, however, I could only see the existence of the bands, but could not see their origin. To observe this, I climbed a summit to the left of a remarkable cleft in the mountain range between the Aiguille du Charmoz and Trelaporte, which strikes most visitors to the Montanvert. The place is that referred to by Professor FORBES at page 84 of the 'Travels,' and which he was prevented from reaching by the fire of stones which a secondary glacier sent down upon him. From the pile of stones on the summit I, however, infer that he or some of his guides must have reached the place. Fig. 23 represents what I observed from this excellent station.

From the base of the Aiguille du Géant and the Periades, a glacier descends which is separated by the Aiguille and promontory of La Noire from the great glacier which descends from the Col du Géant. A small moraine is formed between both, beside which the letter *a* stands in the diagram.   The glacier descending from the Col is bounded on the west by the small moraine *b*, and between *b* and the side of the valley is another little glacier derived from one of the lateral tributaries.

Fig. 23.

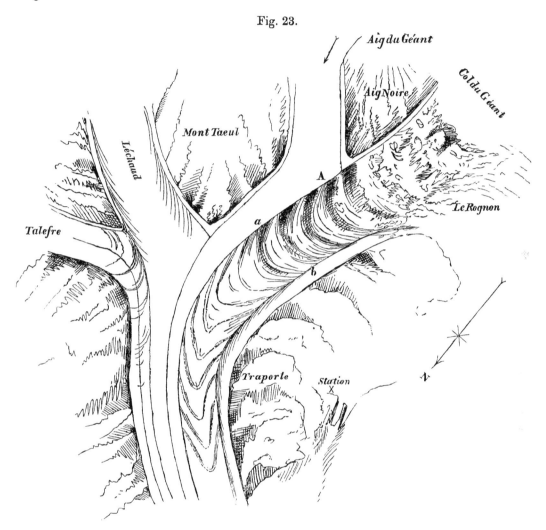

With regard to the " dirt-bands," the following significant fact at once revealed itself. *The dirt-bands extended over that portion of the Glacier du Géant only which lay between the moraines a and b*, or, in other words, were confined to the ice which had descended the great cascade between Le Rognon and La Noire.   It was perfectly evident that the cascade was in some way the cause of the bands.

The description which I have already given of the ice-fall of the Rhone and of the Strahleck arm of the Lower Grindelwald glacier, applies generally to the fall of the Glacier du Géant.   The terraces, however, are here larger, and the protuberances at the base of the fall of grander proportions.   These latter are best seen from a point near A upon the Glacier du Géant; they are steepest on that side, in consequence of the oblique thrust of the western tributaries of the glacier.   All that I have said regarding the toning down of the ridges to rounded undulations which sweep in curves across the glacier, applies here also. Referring to the section of the glacier of the Rhone in fig. 4, it will be seen that the word "dirt" is written opposite to each hollow. In fact the depressions between the protuberances are, to some extent, the collectors of the fine superficial dirt. This is also the case upon the Glacier du Géant; but here I noticed that the *frontal slopes* of the protuberances were also covered with a fine brown mud. Lower down the glacier the swellings disappear, but the dirt retains its position upon the ice, and afterwards constitutes the *dirt bands* of the Mer de Glace.

A remarkable change in the form of the bands occurs where the glacier is forced through the neck of the valley at Trelaporte. They sweep across the Glacier du Géant in gentle curves with their convexity downwards; but in passing Trelaporte the arms of the curves are squeezed more closely together, the vertices are pushed sharply forwards, so that on the whole the bands resemble a series of hyperbolas which tend to coincide with their asymptotes.

Looking down from the Convercle upon the Glacier du Talèfre, a series of swellings like those upon the Glacier du Géant are observed. Along the intervening hollows streams run, and sand and dirt are collected, forming the rudiments, so to speak, of a series of dirt-bands; but these latter never attain anything like the precision of those upon the Mer de Glace. I saw no such bands upon the Léchaud, for here the necessary ice-fall is absent: if bands at all exist on this glacier, they must, I imagine, be of a very rudimentary and defective character.

I will not occupy the time of the Society in describing my various expeditions up the Glacier du Géant in connexion with these bands; but one circumstance, to which the definite printing of the bands is mainly due, must be mentioned. The Glacier du Géant lies nearly north and south, being only 14 degrees east of the true north. Standing with his back to the Col du Géant, an observer looks northward, and consequently the frontal slopes of the protuberances to which I have referred have a *northern aspect*. They therefore retain the snow upon them long after it has been melted from the general surface of the glacier. The summer of 1857 was unusually warm in the Alps, but its great heat was not sufficient entirely to remove the snow. No doubt, in colder summers, the snow is retained upon the slopes all the year round. Now this snow becomes the collector of a fine brown mud, which is scattered over the surface of the glacier. It catches the substance transported by the little rills and retains it. The edges of the snow still remaining, when I was on the glacier, were exceedingly black and dirty; and in many cases the entire surface of the snow appeared as if fine peat mould had been strewn over it. Lower down the glacier this snow melts, but it leaves its sediment behind it, and to this sediment the distinctness of the dirt-bands of the Mer de Glace is mainly due.

The regularity of the bands depends on the regularity with which the glacier is broken, and the ridges or terraces formed as it passes over the brow of the fall. It is the toning down of these ridges which produces the undulations, which are to some extent modified by the squeezing at the base of the fall; and it is the undulations which produce the bands. Thus the latter connect themselves with the transverse fracture of the glacier as it crosses the brow of the fall.

In the figure I have given the general aspect of the bands, but not their number. Thirteen of them exist on the Glacier du Géant. I may add that the bearing I have assigned to this glacier differs from that assigned to it on the map which accompanies Professor FORBES's 'Travels on the Alps,' and which I had with me at the Montanvert. The reason is, that on the map the true north is drawn on the wrong side of the magnetic north, thus making the "Declination" easterly instead of westerly. I have since learned that this error is corrected in the smaller work of Professor FORBES.

It has been affirmed that the dirt-bands cross some of the medial moraines of the Mer de Glace, and they are thus drawn upon the map of Professor FORBES. Were this correct, my explanation would be untenable; but the fact is, that the bands are confined to the Glacier du Géant from beginning to end.

Royal Institution, February 1859.

XV. *On the Structure and Motion of Glaciers.*
By John Tyndall, *F.R.S., Professor of Natural Philosophy, Royal Institution; and*
Thomas H. Huxley, *F.R.S., Fullerian Professor of Physiology, Royal Institution.*

Received and Read, January 15, 1857.

§ 1.

In a lecture given at the Royal Institution on the 6th of June, by Mr. Tyndall, 1856, certain views regarding the origin of slaty cleavage were brought forward, and afterwards reported in the 'Proceedings' of the Institution. A short time subsequently, the attention of the lecturer was drawn by Mr. Huxley to the observations of Professor J. D. Forbes on the veined or laminar structure of glacier ice, and the surmise was expressed, that the same explanation might apply to it as to slaty cleavage. On consulting the observations referred to, the probability of the surmise seemed apparent, and the result was a mutual arrangement to visit some of the Swiss glaciers, for the purpose of observing the structure of the ice. This arrangement was carried out, the field of observation comprising the glaciers of Grindelwald, the Aar, and the Rhone. After returning to England, the one in whose department it more immediately lay, followed up the inquiry, which gradually expanded, until at length it touched the main divisions of the problem of glacier structure and motion. An account of the experiments and observations, and our joint reflections on them, are embodied in the memoir now submitted to the Royal Society.

§ 2. *On the Viscous Theory of Glaciers.*

A glacier is a mass of ice which, connected at its upper extremity with the snow which fills vast mountain basins, thrusts its lower extremity into the warm air which lies below the snow-line. The glacier moves. It yields in conformity with the sinuosities of its walls, and otherwise accommodates itself to the inequalities of the valley which it fills. It is not therefore surprising that the glacier should have been regarded as an ice-river by those who dwelt in its vicinity, or that this notion should have found a place in the speculations of writers upon the subject. The statements of M. Rendu in connexion with this point are particularly distinct:—"There are," he writes, "a multitude of facts which seem to necessitate the belief that the substance of glaciers enjoys a kind of ductility which permits it to model itself on the locality which it occupies, to become thin and narrow, and to elongate itself like a soft paste *." But this observer put forward his speculations with great caution, and often in the form of questions which he confessed his inability to answer. "M. Rendu," says Pro-

* Théorie des Glaciers de la Savoie, p. 84.

fessor FORBES, "has the candour not to treat his ingenious speculations as leading to any certain result, not being founded on experiments worthy of confidence...... My theory of glacial motion, then, is this:—A GLACIER IS AN IMPERFECT FLUID OR VISCOUS BODY, WHICH IS URGED DOWN SLOPES OF A CERTAIN INCLINATION BY THE MUTUAL PRESSURE OF ITS PARTS."

"The sort of consistency to which we refer," proceeds Professor FORBES, "may be illustrated by that of moderately thick mortar, or the contents of a tar-barrel poured into a sloping channel." Treacle and honey are also referred to as illustrative of the consistency of a glacier. The author of the theory endeavours, with much ability, to show that the notion of semifluidity, as applied to ice, is not an absurdity, but on the contrary, that the motion of a glacier exactly resembles that of a viscous body. Like the latter, he urges, it accommodates itself to the twistings of valleys, and moves through narrow gorges. Like a viscous mass, it moves quickest at its centre, the body there being most free from the retarding influence of the lateral walls. He refers to the " Dirt-Bands" upon the surface of the glacier, and shows that they resemble what would be formed on the surface of a sluggish river. In short, the analogies are put forth so clearly, so ably, and so persistently, that it is not surprising that this theory stands at present without a competitor. The phenomena, indeed, are really such as to render it difficult to abstain from forming some such opinion as to their cause. The resemblance of many glaciers to " a pail of thickish mortar poured out;" the gradual changing of a straight line transverse to the glacier into a curve, in consequence of the swifter motion of the centre; the bent grooves upon the surface; the disposition of the dirt; the contortions of the ice, a specimen of which, as sketched near the Heisseplatte upon the Lower Grindelwald glacier, is given in fig. 1, and of which other stri-

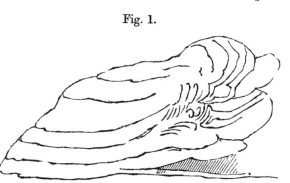

Fig. 1.

king examples have been adduced by M. ESCHER, in proof of the plasticity of the substance,—are all calculated to establish the conviction, that the mass must be either viscous, *or endowed with some other property mechanically equivalent to viscosity*. The question then occurs, is the viscosity real or apparent? Does any property equivalent to viscosity exist, in virtue of which ice can move and mould itself in the manner indicated, and which is still in harmony with our experience of the non-viscous character of the substance? If such a property can be shown to exist, the choice will rest between a quality which ice is *proved* to possess, and one which, in opposition to general experience, it is assumed to possess, in accounting for a series of phenomena which either the real or the hypothetical property might be sufficient to produce. In the next section, the existence of a true cause will be pointed out, which reconciles the properties of ice, exhibited even by hand specimens, with the apparent evidences of viscosity already referred to, and which, though it has been overlooked hitherto, must play a part of the highest importance in the phenomena of the glacier world.

§ 3. *On the Regelation of Ice, and its application to Glacial Phenomena.*

In a lecture given by Mr. Faraday at the Royal Institution on the 7th of June, 1850, and briefly reported in the 'Athenæum' and 'Literary Gazette' for the same month, it was shown that when two pieces of ice, at 32° Fahr., with moistened surfaces, were placed in contact, they became cemented together by the freezing of the film of water between them. When the ice was below 32°, and therefore dry, no adhesion took place between the pieces. Mr. Faraday referred, in illustration of this point, to the well-known experiment of making a snowball. In frosty weather the dry particles of ice will scarcely cohere, but when the snow is in a thawing condition, it may be squeezed into a hard compact mass. On one of the warmest days of last July, when the thermometer stood at upwards of 80° Fahr. in the shade and above 100° in the sun, a pile of ice-blocks was observed by one of us in a shop window, and he thought it interesting to examine whether the pieces were united at their places of contact. Laying hold of the topmost block, the whole heap, consisting of several large lumps, was lifted bodily out of its vessel. Even at this high temperature the pieces were frozen together at the places of contact, though the ice all round these places had been melted away, leaving the lumps in some cases united by slender cylinders of the substance. A similar experiment may be made in water as hot as the hands can bear; two pieces of ice will freeze together, and sometimes continue so frozen in the hot water, until, as in the case above mentioned, the melting of the ice around the points of contact leaves the pieces united by slender columns of the substance.

Acquainted with these facts, the thought arose of examining how far, in virtue of the property referred to, the *form* of ice could be changed without final prejudice to its continuity. It was supposed that though crushed by great pressure, new attachments would be formed by the cementing, through regelation, of the severed surfaces; and that a resemblance to an effect due to viscosity might be produced. To test this conjecture the following experiments were made:—Two pieces of seasoned boxwood, A and B, fig. 2, 4 inches square and 2 deep, had two cavities hollowed out, so that when one was placed upon the other, a lenticular space, shown in section at C, was enclosed between them. A *sphere* of compact, transparent ice, of a volume rather more than sufficient to fill the cavity, was placed between the pieces of wood, and subjected to the pressure of a small hydraulic press. The ice broke, as was expected, but it soon re-attached itself; the pressure was continued, and in a few seconds *the sphere was*

Fig. 2.

reduced to a transparent lens of the shape and size of the mould in which it had been formed.

This lens was placed in a cylindrical cavity, two inches wide and half an inch deep, hollowed out in a piece of boxwood, C, fig. 3, as before; a *flat* plate, D, of the wood being placed over the lens, it was submitted to pressure. The lens broke as the sphere did, but the fragments attached themselves in accordance with their new conditions, *and in less than half a minute the mass was taken from the mould a transparent cake of ice.*

Fig. 3.

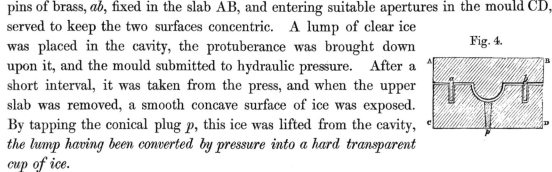

The substance was subjected to a still severer test. A hemispherical cavity was hollowed out in a block of boxwood, and a protuberant hemisphere was turned upon a second slab of the wood, so that, when the protuberance and the cavity were concentric, a distance of a quarter of an inch separated the convex surface of the former from the concave surface of the latter. Fig. 4 shows the arrangement in section. The pins of brass, *ab*, fixed in the slab AB, and entering suitable apertures in the mould CD, served to keep the two surfaces concentric. A lump of clear ice was placed in the cavity, the protuberance was brought down upon it, and the mould submitted to hydraulic pressure. After a short interval, it was taken from the press, and when the upper slab was removed, a smooth concave surface of ice was exposed. By tapping the conical plug *p*, this ice was lifted from the cavity, *the lump having been converted by pressure into a hard transparent cup of ice.*

Fig. 4.

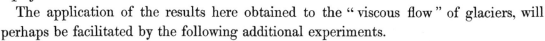

The application of the results here obtained to the " viscous flow " of glaciers, will perhaps be facilitated by the following additional experiments.

A block of boxwood (A, fig. 5), 4 inches long, 3 wide and 3 deep, had its upper surface slightly curved, and a longitudinal groove (shown in dots in the figure), an inch wide and an inch deep, worked into it. A slab of the wood was prepared, the under surface of which was that of a convex cylinder, curved to the same degree as the concave surface of the former piece. The arrangement is shown in section at B. A straight prism of clear ice, 4 inches in length, an inch wide, and a little more than an inch in depth, was placed in the groove, and the upper slab of boxwood was placed upon it. The mould was submitted to hydraulic pressure, as in the former cases; the prism broke as a matter of course, but the quantity of ice being rather more than sufficient to fill the groove, and hence projecting above its edge, the pressure brought the fragments together and re-established the continuity of the ice. After a few seconds it was taken from the mould, bent as if it had been a plastic mass. Three other moulds similar to the last, but of augmenting curvature, were afterwards made use of, the same prism being passed through all of them in succession. *At the*

Fig. 5.

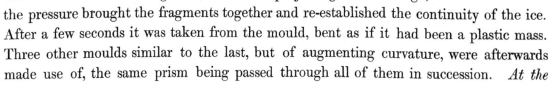

conclusion of the experiments the prism came out, bent to a transparent semi-ring of solid ice.

In this way, by the proper application of force, all the bendings and contortions observed in glacier ice, and adduced in proof of its viscosity, can be accurately imitated. Any observer, seeing a straight bar of ice converted into a continuous semi-ring without being aware of the quality referred to, and having his attention fixed on the changes of external form alone, would be naturally led to the conclusion that the substance is viscous. But it is plainly not viscosity, properly so called, which enables it to change its shape in this way, but a property which has hitherto been entirely overlooked by writers upon glaciers.

It has been established by observation, that a vertical layer of ice originally plane, and perpendicular to the axis of a glacier, becomes bent, because the motion of its ends is retarded in comparison with that of its centre. This is the fact upon which the viscous theory principally rests.

In the experiments with the straight prism of ice, four successive moulds, gradually augmenting in curvature, were made use of. In passing suddenly from the shape of one to that of the other, the ice was fractured, but the pressure brought the separated surfaces again into contact and caused them to freeze together, thus restoring the continuity of the mass. The fracture was in every case both audible and tangible; it could be heard and it could be felt. A series of cracks occurred in succession as the different parts of the ice-prism gave way, and towards the conclusion of the experiment, the crackling in some instances melted into an almost musical tone. But if instead of causing the change to take place by such wide steps as those indicated; if instead of four moulds, forty, or four hundred were made use of; or better still, suppose a single mould to have the power of gradually changing its curvature from a straight line to a semicircle under the hydraulic press; the change in the curvature of the ice would closely approximate to that of a truly plastic or viscous body. This represents the state of things in a glacier. A transverse plate of ice, situated between the mass in front of it and the mass behind, is virtually squeezed in a press of the description which has been just imagined. The curvature of the ice-mould *does* change in the manner indicated, and so slowly, that the bending closely resembles what would take place if the substance were viscous. The gradual nature of the change of curvature may be inferred from an experiment made by Professor FORBES on an apparently compact portion of the Mer de Glace. He divided a distance of 90 feet transverse to the axis of the glacier into spaces of two feet each, and observed with a theodolite the gradual passage of this straight line into a bent one. The row of pins bent gradually so as to form a curve convex towards the lower extremity of the glacier; their deviations from a perfect curve were slight and irregular, nor was any great dislocation to be observed throughout their whole extent. After six days the summit of the curve formed by the forty-five pins was one inch in advance of the straight chord which united its two ends. It is not surprising if, with this extremely gradual change, the motion should have appeared to be the result of viscosity. It may, however, be remarked, that the slight and irregular variations to which Professor FORBES alludes, and which are such as would occur if the motion were such as we suppose it to be, are likely to throw much light upon the problem. It is also extremely probable that the motion, if effected in the manner referred to, will be sometimes accompanied by an audible crackling of the mass. To this we paid but little attention when on the ground; for the significance of this as well as of many other points was first suggested by the experiments made after our return. It is, however, we believe, a phenomenon of common occurrence. Professor FORBES calls the glacier a " crackling mass;" he speaks of the ice " cracking and straining forwards;" and in that concluding passage of his 'Travels' which has excited such general admiration, he says of the glacier, "it yields groaning to its fate." Other observers make use of similar expressions. M. DESOR also speaks of the sudden change of the colour of the blue veins of the ice where a portion of the central moraine near the Abschwung is cleared away; the observation is very remarkable. " Au moment," says M. DESOR, " où on la met à découvert, la glace des bandes bleues est parfaitement transparente, l'œil y plonge jusqu'à une profondeur de plusieurs pieds, mais cette pureté ne dure qu'un instant, et l'on voit bientôt se former des petites fêlures d'abord superficielles, qui se combinent en réseau de manière à enlever peu à peu à la glace bleue toute sa transparence. Ces fêlures propagent également dans les bandes blanches, et lorsqu'on approche l'oreille de la surface de la glace, en entend distinctement un *leger bruit de crépitation* qui les accompagnent au moment de leur formation." These facts appear to be totally at variance with the idea of viscosity.

In a chapter on the " Appearance of the larger Glaciers," in an interesting little work by M. MOUSSON of Zürich, for which one of us has to thank the kindness of Professor CLAUSIUS, the phenomena which they exhibit are thus described[*]:—" The appearance of a large glacier of the first order has been compared, not without reason, with that of a high swelled, and suddenly solidified stream. It winds itself in a similar manner through the curving of the valley, is deflected by obstacles, contracts its width, or spreads itself out...... In short, the form is modified in the most complete manner to suit the character and irregularities of its bed. To this capacity to change its form, the ice of glaciers unites another property, which reminds us of the fluid condition; namely, the capability of joining and blending with other ice. Thus we see separate glacier branches perfectly uniting themselves to a single trunk; regenerated glaciers formed from crushed fragments; fissures and chasms closed up, and other similar appearances. These phenomena evidently point to a slow movement of the particles of which the glacier consists; strange as the application of such an idea to a solid brittle mass such as glacier ice may appear to be. The solution of this enigma constitutes one of the most difficult points in the explanation of glaciers."

When the appearances here enumerated are considered with reference to the experiments on the regelation of ice above described, the enigma referred to by the writer appears to have received a satisfactory solution. The glacial valley is a mould through

* Die Gletscher des Jetztzeit, by ALBERT MOUSSON. Zürich, 1854.

which the ice is pressed by its own gravity, and to which it will accommodate itself, while preserving its general continuity, as the hand specimens do to the moulds made use of in the experiments. Two glacial branches unite to form a single trunk, by the regelation of their pressed surfaces of junction. Crevasses are cemented for the same reason; and the broken ice of a cascade is reconstituted, as a heap of fragments under pressure become consolidated to a single mass. To those who occupy themselves with the external conditions merely of a glacier, it may appear of little consequence whether the flexures exhibited by the ice be the result of viscosity or of the principle demonstrated by the experiments above described. But the natural philosopher, whose vocation it is to inquire into the inner mechanism concerned in the production of the phenomena, will discern in the yielding of a glacier a case of simulated fluidity hitherto unexplained, and perhaps without a parallel in nature.

§ 4. *On the Veined Structure of Glacier Ice.*

This structure has been indifferently called the "veined structure," the "banded structure," the "ribboned structure," and the "laminar structure" of glacier ice. In a communication to the Geological Society of France assembled at Porrentruy in September 1838, M. GUYOT gave the following interesting description of the phenomenon:—" Since the word layer has escaped me, I cannot help recording as a subject of investigation for future observers a fact, regarding which I dare not hazard an explanation; especially as I have not encountered it more than once. It was at the summit of the Gries, at a height of about 7500 feet, a little below the line of the first or high nevé, where the ice passes into a state of granular snow..... In ascending to the origin of this latter (the glacier of Bettelmatten), for the purpose of examining the formation and direction of the great transverse fissures, I saw under my feet the surface of the glacier entirely covered with regular furrows, from 1 to 2 inches in width, hollowed in a half snowy mass, and separated by protruding plates of an ice more hard and transparent. It was evident that the mass of the glacier was here composed of two sorts of ice, one that of the furrows, still snowy and more easily melted, the other that of the plates, more perfect, crystalline, glassy and resistent; and that it was to the unequal resistance which they presented to the action of the atmosphere that was due the hollowing of the furrows and the protrusion of the harder plates. After having followed them for several hundred yards, I reached the edge of a great fissure, 20 or 30 feet wide; which cutting the plates and furrows perpendicularly to their direction, and exposing the interior of the glacier to a depth of 30 or 40 feet, permitted the structure to be observed on a beautiful transverse section. As far down as my vision could reach I saw the mass of the glacier composed of a multitude of layers of snowy ice, each two separated by one of the plates of ice of which I have spoken, and forming a whole regularly laminated in the manner of certain calcareous slates."

A description of this structure, as observed upon the glacier of the Aar, was communicated by Professor FORBES to the Royal Society of Edinburgh on the 6th of De-

cember 1841, and published in the Edinburgh New Philosophical Journal for 1842*.
He was undoubtedly the first to give the phenomenon a theoretic significance.

While engaged in the Lower Grindelwald glacier, we separated plates of ice perpendicular to the lamination of the glacier. The appearance presented on looking through
them, was that sketched in fig. 6. The layers of transparent ice
seemed imbedded in a general milky mass; through the former
the light reached the eyes, while it was intercepted by the latter.
Some of the transparent portions were sharply defined, and exhibited elongated oval sections, resembling that of a double convex lens, and we therefore called this disposition of the veins
" *the lenticular structure.*" In other cases, however, the sharpness of outline did not exist, but still the tendency to the lenticular form could be discerned, the veins in some cases terminating
in washy streaks of blue. This structure is probably the same as
that observed by Professor FORBES on the Glacier des Bossons,
and described in the following words:—" The veins and bands .... are not formed
in this glacier by a simple alternation of parallel layers, but the icy bands have all the
appearance of posterior infiltration, occasioned by fissures, *thinning off both ways*†."

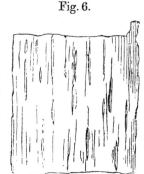

Fig. 6.

In 1842 Professor FORBES undertook the survey and examination of the Mer de Glace,
and finally arrived at a theory of glacier lamination, which both in his 'Travels' and in a
series of letters, extending over a period of several years, he has expounded and illustrated with great skill. The theory is summed up in the following words:—" The
whole phenomena in the case of any of the semifluids I have mentioned (treacle, tar,
&c.), are such as, combined with the evidence which I have given, that the motion of a
glacier is actually such as I have described that of a viscid fluid to be, can leave, I
think, no reasonable doubt, *that the crevices formed by the forced separation of a half
rigid mass, whose parts are compelled to move with different velocities, becoming infiltrated
with water, and frozen during winter, produce the bands which we have described* ‡."

This theory has been opposed by Mr. HOPKINS, whose excellent papers, published in
the 26th volume of the Philosophical Magazine, are replete with instruction as to the
mechanical conditions of glaciers. On the other hand, the theory of Professor FORBES
is defended in the same journal by Dr. WHEWELL §. We will leave the points discussed

---

* This communication gave rise to a discussion as to priority between Professor FORBES and M. AGASSIZ,
for the details of which we must refer to the original papers on the subject.

† Travels, p. 181.

‡ Ibid. p. 377. M. AGASSIZ also seems disposed to regard the blue bands as the result of the freezing
up of fissures, which, however, are supposed to be formed in a manner different from that assumed by Professor FORBES. But M. AGASSIZ calls the attention of future observers to some of the related phenomena;
and gives it as his opinion, "qu'il n'est aucune phénomène dont l'explication offre plus des difficultés."
See his important work, 'Système Glacière,' which, until quite recently, we had not the opportunity of
examining.

§ Philosophical Magazine, S. 3. vol. xxvi. pp. 171, 217.

in their communications for the present untouched, and confine ourselves to stating a few of the circumstances which appear to us to render the theory doubtful.

1. It is not certain that the colds of winter penetrate to depths sufficient to produce the blue veins, which, it is affirmed, are "an integral part of the inmost structure" of the ice. SAUSSURE was of opinion that the frosts of winter did not penetrate to a greater depth than 10 feet, even at the summit of Mont Blanc, and Professor FORBES considers this opinion to be a just one. But if so, there would be some difficulty in referring to the frosts of winter the blue veins which M. AGASSIZ observed at a depth of 120 feet below the surface of the glacier of the Aar.

2. It will be remembered that M. GUYOT's statement regarding the blue veins is, that he saw the mass of the glacier composed of a multitude of layers of white ice, separated, each from the other, by a plate of transparent ice. The description of Professor FORBES is briefly this:—" Laminæ or thin plates of transparent blue ice, alternate in most parts of every glacier with laminæ of ice, not less hard and perfect, but filled with countless air-bubbles which give it a frothy semitransparent look." But there is another form of the blue veins, already referred to, which consists in transparent lenticular masses imbedded in the general substance of the white ice. Horizontal sections of these transparent lenses were exposed upon the surface of the Grindelwald glacier, and vertical sections of them upon the perpendicular sides of the water-courses, and upon the walls of the crevasses. The following measurements, taken on the spot, will give an idea of their varying dimensions:—

Fig. 7.

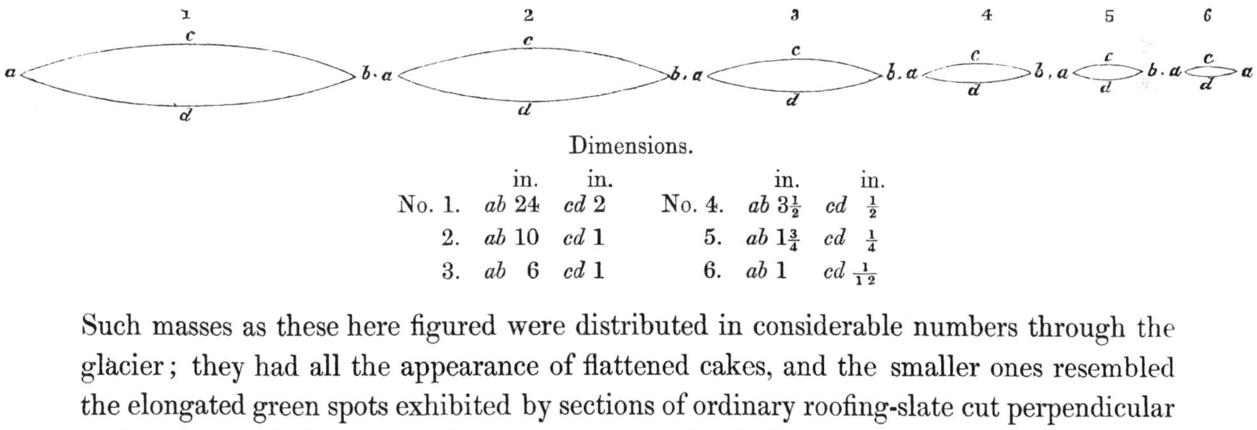

Dimensions.

	in.	in.		in.	in.
No. 1.	ab 24	cd 2	No. 4.	ab $3\frac{1}{2}$	cd $\frac{1}{2}$
2.	ab 10	cd 1	5.	ab $1\frac{3}{4}$	cd $\frac{1}{4}$
3.	ab 6	cd 1	6.	ab 1	cd $\frac{1}{12}$

Such masses as these here figured were distributed in considerable numbers through the glacier; they had all the appearance of flattened cakes, and the smaller ones resembled the elongated green spots exhibited by sections of ordinary roofing-slate cut perpendicular to the planes of cleavage. Now it appears mechanically impossible that a solution of continuity, such as that supposed, could take the form of the detached lenticular spaces above figured.

3. The fissures to which the blue veins owe their existence are stated to be due to the motion of the glacier; and as this motion takes place both in summer and winter, it is to be inferred that the fissures are produced at both seasons of the year. Now as the fissures formed in winter cannot be filled with ice during that season for want of *water*, and as those formed in the ensuing summer cannot, while summer continues, be

frozen for want of *cold*, we ought at the end of each summer to have *a whole year's fissures* in the ice. These fissures, which the ensuing winter is, according to the theory, to fill with blue ice, must, in summer, be filled with *blue water*. *Why then are they not seen in summer?* The fissures are such as can produce plates of ice varying "from a small fraction of an inch to several inches in thickness," which, according to our own observations, produce lenticular masses of ice 2 feet long and 2 inches thick, or even (for we have seen pieces of this description) 10 feet long and 10 inches thick; and M. DESOR informs us in the memoir from which we have already quoted, that under the medial moraine of the Aar glacier, there are bands 10 inches and even a foot in thickness. Such fissures could not escape observation if they existed, but they never have been observed, and hence the theory which makes their pre-existence necessary to the production of the blue veins appears to us improbable.

§ 5. *On the Relation of Slaty Cleavage to the Veined Structure.*

Within the last few years a mechanical theory of the cleavage of slate rocks has been gradually gaining ground among those who have reflected upon the subject. The observations of the late DANIEL SHARPE appear to have originated this theory. He found that fossils contained in slate rocks were distorted in a manner which proved that they had suffered compression in a direction at right angles to the planes of cleavage. His specimens of shells, which are preserved in the Museum of Practical Geology, and other compressed fossils in the same collection, illustrate in a remarkable manner his important observations. The subsequent microscopic observations of Mr. SORBY, carried out with so much skill and patience, show convincingly that the effects of compression may be traced to the minutest constituents of the rocks in which cleavage is developed. More recently, Professor HAUGHTON has endeavoured to give numerical accuracy to this theory, by computing, from the amount of the distortion of fossils, the magnitude of the change which cleaved rocks have undergone. By the united testimony of these and other observers, whose researches have been carried out in different places, the association of cleavage and compression has been established in the most unequivocal manner; and hence the question naturally arises, "Is the pressure sufficient to produce the cleavage?" SHARPE appears to have despaired of an experimental answer to this question. "If," says he, "to this conclusion it should be objected, that no similar results can be produced by experiment, I reply, that we have never tried the experiment with a power at all to be compared with that employed; and that this may be one of the many cases where our attempts to imitate the operations of nature fail, owing to the feebleness of our means, and the shortness of the period during which we can employ them." The same opinion appears to have been entertained by Professor FORBES:—"The experiment," he says, "is one which the boldest philosopher would be puzzled to repeat in his laboratory; it probably requires acres for its scope, and years for its accomplishment."

While one of us was engaged in 1855 in examining the influence of pressure upon magnetism, he was fortunate enough to discover that in white wax, and other bodies, a cleavage of surpassing fineness may be developed by pressure, and he afterwards endeavoured, in a short paper*, to show the application of this result, both to slaty cleavage and to a number of other apparently unrelated phenomena. The theory propounded in this paper may be thus briefly stated. If a piece of clay, wax, marble or iron be broken, the surface of fracture will not be a plane surface, nor will it be a surface dependent only on the form of the body and the strain to which it has been subjected; the fracture will be composed of innumerable indentations, or small facets, each of which marks a surface of weak cohesion. The body has yielded, where it could yield, most easily, and in exposing these facets, in some cases crystalline, in others purely mechanical, wherever the mass is broken, it is shown to be composed of an aggregate of irregularly-shaped parts, which are separated from each other by surfaces of weak cohesion. Such a quality must, in an eminent degree, have been possessed by the mud of which slate-rocks are composed, after the water with which the mud had at first been saturated had drained away; and the result of the application of pressure to such a mass would be, to develope in it a lamination similar to that so perfectly produced on a small scale in white wax. Thus one cause of cleavage may be stated, in general terms, to be the conversion by pressure of irregularly-formed surfaces of weak cohesion into parallel planes. To produce lamination in a compact body such as wax, it is manifest that while it yields to the compression in one direction, it must have an opportunity of expanding in a direction at right angles to that in which the pressure is exerted; a second cause is the lateral sliding of the particles which thus takes place, and which may be very influential in producing the cleavage†.

Before attempting to show the connexion between this theory and the case at present under consideration, a mode of experiment may be described which was found to assist in forming a conception of the mechanical conditions of a glacier, and which has already been resorted to by Professor FORBES in demonstration of the viscous theory. Owing to

---

* Proceedings of the Royal Institution, June 1856; Philosophical Magazine for July 1856.

† Three principal causes may operate in producing cleavage:—1st, the reducing of surfaces of weak cohesion to parallel planes; 2nd, the flattening of minute cavities; and 3rd, the weakening of cohesion by tangential action. The third action is exemplified by the state of the rails near a station where the break is applied. In this case, while the weight of the train presses vertically, its motion tends to cause longitudinal sliding of the particles of the rail. Tangential action does not however necessarily imply a force of the latter kind. When a solid cylinder, an inch in height, is squeezed by vertical pressure to a cake a quarter of an inch in height, it is impossible, physically speaking, that the particles situated in the same vertical line shall move laterally with the same velocity; but if they do not, the cohesion between them will be weakened or ruptured. The pressure will produce new contact, and if the new contact have a cohesive value equal to that of the old, no cleavage from this cause can arise. The relative capacities of different substances for cleavage, appears to depend in a great measure upon their different properties in this respect. In butter, for example, the new attachments are equal, or nearly so, to the old, and the cleavage is consequently indistinct; in wax this does not appear to be the case, and hence may arise in a great degree the perfection of its cleavage. The further examination of this subject promises interesting results.

the property of ice described in § 3, the resemblance between the motion of a substance like mud and that of a glacier is so great, that considerable insight regarding the deportment of the latter may be derived from a study of the former. From the manner in which mud yields when subjected to mechanical strain, we may infer the manner in which ice would be *solicited to yield* under the same circumstances.

To represent then the principal accidents of a glacial valley, a wooden trough, ABCD, fig. 8, of varying width and inclination, was made use of. From A to C the

Fig. 8.

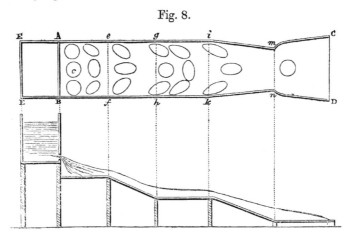

trough measures 6 feet, and from A to B, 15 inches. It is divided into five segments; that between AB and *ef* is level, or nearly so, that between *ef* and *gh* is inclined; from *gh* to *ik* is again nearly level; from *ik* to *mn* inclined, while from *mn* to CD the inclination is less than between *ik* and *mn*. The section of the bottom of the trough is figured underneath the plan. ABEF is a box supported at the end of the trough, and filled with a mixture of water and fine pipe-clay. The front, AB, can be raised, like a sluice, and the mud permitted to flow regularly into the trough. While the mud is in slow motion, a coloured circle, *c*, is stamped upon the white clay between AB and *ef*; the changes of shape which this circle undergoes in its passage downwards will indicate the forces acting upon it. The circle first moves on, being rather compressed, in the direction of the length of the trough until it reaches *ef*, on crossing which, and passing down the subsequent slope, it elongates as in the figure. Between *gh* and *ik* the figure passes through the circular form, and assumes that of an ellipse, whose shorter axis is parallel to the length of the trough. It is manifest from this that the mud between *ef* and *gh* is in a state of longitudinal tension, while between *gh* and *ik* its state is that of longitudinal compression. On crossing *ik* and descending the second incline, the figure is again drawn out longitudinally, while between *mn* and CD the ellipse widens on account of the permission given to lateral expansion by the augmented width of the trough.

The side circles in the same figure will enable us to study the influence of lateral friction upon the descending stream. These circles are distorted into ellipses, whose major axes are oblique to the direction of the trough's length. Above the line *ef* central fissures perpendicular to the axis of the trough cannot be formed; for here, instead of tending to open into fissures, the flattening of the central circle shows that the mud is longitudinally compressed. On the slope below *ef*, the distortion of the circles into ellipses is very pronounced; and as the longer axis of each ellipse marks the line of maximum tension, and as the tendency of the mass is to form a fissure at right angles to such a line, we should have here, if the substance were not so plastic as to prevent the formation of fissures, the state of things observed upon the corresponding portion of the glacier; namely, central fissures perpendicular to the longitudinal axis of the trough, and side fissures inclined to the same axis because pointing in the direction of the shorter axis of each ellipse. Between *gh* and *ik* the longitudinal tension is changed to compression; the central figure is flattened, while the side ones remain stretched. In the corresponding portion of the glacier we should expect the central fissures formed between *ef* and *gh* to be squeezed together and closed up, while the lateral ones would remain open. This is also the case*. Between *ik* and *mn* we have again longitudinal tension, and at the corresponding portions of the glacier the transverse central crevasses ought to reappear, which they actually do. Below the line corresponding to *mn*, the widening of the valley, in the case now in our recollection, causes the ridges produced at the previous slope to break across and form prismatic blocks; while lower down the valley these prisms are converted by the action of sun and rain into shining minarets of ice. These results appear to be in perfect accordance with those arrived at by Mr. HOPKINS on strict mechanical reasoning†.

We will now seek to show the analogy of slaty cleavage to the laminar structure of glacier ice. Referring to fig. 8, it will be seen that in the distortion of the side circles one diameter is elongated to form the transverse axis of the ellipse, while another is compressed to form the conjugate axis. In a substance like mud, as the elongation of the major axis continues, its inclination to the axis of the glacier continually changes; but were the substance one of limited extensibility like ice, fissures would be formed when the tension had reached a sufficient amount, or in other words, when the major axis of the ellipse had assumed a definite inclination to the axis of the glacier.

Thus, in a glacier of the form represented by our trough, owing to the swifter motion

---

* The possibility of the coexistence of lateral crevasses and compression at the centre may, perhaps, be thus rendered manifest :—let *ab*, *cd* be two linear elements of a glacier, situated near its side S I.

Suppose, on passing downward, the line *ab* becomes shortened by longitudinal pressure to *a'b'*, and *cd* to *c'd'*, which latter has passed *a'b'* on account of its greater distance from the side of the glacier. Taking the figure to represent the true change both of dimension and position, it is plain, that though each element has been *compressed*, the differential motion has been such as to *distend* the line of particles joining *a* and *d*, in the ratio $\frac{ad}{a'd'}$. If this ratio be more than that which the extensibility of ice can permit of, a side fissure will be formed.

† Philosophical Magazine, 1845, vol. xxvi.

of the centre, we have a line of maximum pressure oblique to the wall of the glacier, and a line of maximum tension perpendicular to the former; crevasses are formed at right angles to the direction of tension, and *it is approximately at right angles to the direction of pressure, as in the case of slate rocks, that the lamination of glacier ice is developed.*

Under ordinary circumstances, therefore, the lamination near the sides of the glacier would, in accordance with the theory of compression, be oblique to the sides, which it actually is. It would be transverse to the crevasses wherever they occur, which it actually is. If the bed of a glacier at any place be so inclined as to cause its central portions to be longitudinally compressed, the lamination, if due to compression, ought to be carried across the glacier at such a place, being transverse to the axis of the glacier at its centre, which is actually the case. This relation of the planes of lamination to the direction of pressure is constant under a great variety of conditions. A local obstacle which produces a thrust and compression is also instrumental in developing the veined structure. In short, so far as our observations reach, wherever the necessary pressure comes into play, the veined structure is developed; being always approximately at right angles to the direction in which the pressure is exerted.

But we will not rely in the present instance upon our own observations alone. Before he formed any theory of the structure, and in his first letter upon the subject, Professor FORBES remarks, that " the whole phenomenon has a good deal the air of a structure induced *perpendicularly to the lines of greatest pressure.*" His later testimony is in substance the same. In his thirteenth letter, read before the Royal Society of Edinburgh on the 2nd of December, 1846, he says that the blue veins are formed *where the pressure is most intense.* In his reference to the development of the laminar structure on the glacier of the Brenva, the pressure is described as being " *violent,*" the effect being such as to produce " *a true cleavage when the ice is broken with a hammer or cut with an axe.*" So also with regard to the glacier of Allalein*, he says " the veined structure is especially developed in front, *i. e.* against the opposing side of the valley, where the pressure is greater than laterally." In fact, the parallelism of the phenomenon to that of slaty cleavage struck Professor FORBES himself, as is evident from the use of the term " now" in the following passage:—" It will be understood that I do not *now* suppose that there is any parallelism between the phenomenon of rocky cleavage and the ribboned structure of the ice." This reads like the giving up of a previously held opinion; the term *now* being printed in italics by Professor FORBES himself. The adoption of the viscous theory appears to have carried the renunciation of this idea in its train.

Later still, and from a source wholly independent of the former, we have received additional testimony on the point in question. The following quotation is from a letter, dated 16th November, 1856, received by one of us from Professor CLAUSIUS of Zurich, so well known in this country through his important memoirs on the Mechanical Theory of Heat:—" I must now," writes M. CLAUSIUS, " describe to you another singular coincidence. I had read your paper upon the cleavage of rocks........ and it occurred to

* Travels, p. 352.

me at the time that the blue veins of glaciers, which indeed I had not seen, but which had been the subject of repeated conversations between Professor STUDER of Berne, Professor ESCHER VON DER LINTH, and myself, might be explained in the same manner. When, therefore, I reached the Rhone glacier for the first time, I walked along it for a considerable extent, and directed my attention particularly to the structure. I repeated this on the other glaciers which I visited during my excursion. I did not indeed pursue the subject so far into detail as to be able in all cases to deduce the blue veins from the existing conditions of pressure, but the correctness of the general explanation impressed itself upon me more and more. This was particularly the case in the glacier of the Rhone, where I saw the blue bands most distinctly, and where also their position harmonized with the pressure endured by the glacier when it was forced to change the direction of its motion. You can therefore imagine how astonished I was to learn that at the same time, and on this very glacier among others, you had been making the same investigations." It ought also to be remarked, that a similar thought occurred to Mr. SORBY, from whom after his return from Switzerland one of us received a note, in which pressure was referred to as the possible cause of the veined structure of glacier ice.

A fine example of ice lamination is that produced by the mutual thrust of two confluent glaciers. The junction of the Lauter Aar and Finster Aar glaciers to form the glaciers of the Unter Aar is a case in point, and the results obtained with a model of this glacier were highly interesting. Fig. 9 is a sketch of the trough in which the experiments were made. The branch terminating at UL is meant to represent the Lauter Aar glacier; that ending at FN the Finster Aar branch. The point at A represents the "Abschwung," so often referred to in the works of M. AGASSIZ. B and B′ are two boxes with sluice fronts, from which the mud flows into the trough. The object was to observe the mechanical state of the mass along the line of junction of the two streams, and along their respective centres, and compare the result with the observations upon

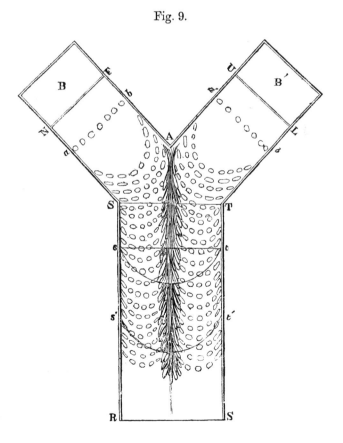

Fig. 9.

the glacier itself. The mud was first permitted to flow simultaneously from both boxes, and after it had covered the bottom of the trough to some distance below the line ST, the end of a glass tube was dipped into a fine mixture of the red oxide of iron and water, and the two arms of the glacier were covered all over with small circles similar to those between the points *ab* and *a'b'*. The mud was then permitted to flow, and the mechanical strains exerted on it were inferred from the distortion of the small circles. The figure represents the result of the experiment. The straight rows of circles bent in the first place into curves; at the point A both streams met, and by their mutual push actually squeezed the circles into lines. Along this central portion in the glacier itself the great medial moraine stands, and under it and beside it, as already stated, the lamination is most strikingly developed; the blue veins being parallel to the axis of the glacier, or, in other words, coinciding with the direction of the central moraine. Midway between the moraine and the sides of the glacier the structure is very imperfectly developed, and the deportment of our model, which shows that the circles here scarcely change their form, tells us that this is the result which ought to be expected. It may be urged, that the structure is here developed, because of the sliding motion produced by the swifter flow of one of the glaciers; but some of the experiments with the model were so arranged, that both of the branch streams flowed with the same velocity; the distortions, however, were such as are shown in the figure. The case is precisely the same in nature. On reference to the map of M. AGASSIZ, we find a straight line set out across the Unter Aar glacier bent in three successive years into a curve; but on the central moraine, which marks the common limit of the constituent streams, we find no breach in the continuity of the curve, which must be the case if one glacier slid past the other.

§ 6. *On the " Dirt-Bands" of Glaciers.*

Wherever the veined structure of a glacier is highly developed, the surface of the ice, owing to the action of the weather, is grooved in accordance with the lamination underneath. These grooves are sometimes as fine as if drawn by a pencil, and bear in many instances a striking resemblance to those produced by the passage of a rake over a graveled surface. In the furrows of the ice the smaller particles of dirt principally rest, and the direction of the furrows, which always corresponds with that of the blue veins, is thus rendered so manifest, that a practised observer can at any moment pronounce upon the direction of the lamination from the mere inspection of the surface of a glacier. But besides these narrow grooves, larger patches of discoloration are sometimes observed, which take the form of curves sufficient in width to cover hundreds or thousands of the smaller ones. To an eye placed at a sufficient height above a glacier on which they exist, their general arrangement and direction are distinctly visible. To these Professor FORBES has given the name of " Dirt-Bands," and the discovery of them, leading as it did to his theories of glacial motion, and of the veined structure of glacial ice, is to be regarded as one of the most important of his observations.

On the evening of the 24th of July he walked up the hill of Charmoz to a height of about 1000 feet above the level of the glacier, and, favoured by the peculiar light of the hour, observed " a series of nearly hyperbolic brownish bands on the glacier, the curves pointing downwards and the two branches mingling indiscriminately with the moraines." The cause of these bands was the next point to be considered, and his examination of them satisfied him " that the particles of earth and sand and disintegrated rock, which the winds and avalanches and water-runs spread over the entire breadth of the ice, formed a *lodgement* in those portions of the glacier where the ice was most porous, and that, consequently, the ' dirt-bands ' were merely *indices of a peculiarly porous veined structure traversing the mass of the glacier in these directions.*"

Professor FORBES was afterwards led to regard these intervals as the marks of the annual growth of the glacier; he called the dirt-bands " annual rings [*]," and calculated, from their distance apart, the yearly rate of movement. In fine, the conclusion which he deduces from the dirt-bands is, that a glacier throughout its entire length is formed of alternate segments of porous and of hard ice. The dirt which falls upon the latter is washed away, as it has no hold upon the surface; that which falls upon the former remains, because the porous mass underneath gives it a lodgement. " The cause of the dazzling whiteness of the glacier des Bossons at Chamouni is the comparative absence of these layers of granular and compact ice: the whole is nearly of uniform consistence, the particles of rock scarcely find a lodgement, and the whole is washed clean by every shower[†]." " It must be owned, however," says Professor FORBES, " that there are several difficulties which require to be removed, as to the recurrence of these porous beds." In his fifteenth letter upon glaciers, and in reference to some interesting observations of Mr. MILWARD'S, he endeavoured to account for the difference of structure by referring it to an annual " gush" of the ice, which is produced by the difference of action in summer and winter. We are ignorant of the nature of the experiments on which this theory of the dirt-bands is founded, and would offer the following simple explanation of those which came under our own observation.

Standing at a point which commanded a view of the Rhone glacier, both above and below the cascade, we observed that the extensive ice-field above was discoloured by sand and débris, distributed without regularity. At the summit of the ice-fall the valley narrows to a gorge, and the slope downwards is for some distance precipitous. In descending, the ice is greatly shattered; in fact, the glacier is broken repeatedly at the summit of the declivity, transverse chasms being thus formed; and these, as the ice descends, are broken up into confused ridges and peaks, with intervening spaces, where the mass is ground to pieces. By this breaking up of the glacier the dirt upon its surface undergoes fresh distribution: instead of being spread uniformly over the slope, spaces are observed quite free from dirt, and other spaces covered with it, but there is no appearance of regularity in this distribution. At some places large

---

[*] " I cannot help thinking that they are the *true annual rings* of the glacier, which mark its age like those of a tree."—Appendix to Travels, p. 408.

[†] Travels, p. 406.

irregular patches appear, and at others elongated spaces covered with dirt. Towards the bottom of the cascade the aspect changes; but still, were the eye not instructed by what it sees lower down, the change would have no significance. When the ice has fairly escaped from the gorge, and has liberty to expand laterally in the valley below, the patches of dirt are squeezed by the push behind them, and drawn laterally into narrow stripes, which run across the glacier; and as the central portion moves more quickly than the sides, these strips of discoloration form curves which turn their convexity downwards, constituting, we suppose, the "Dirt-Bands" of Professor FORBES. On the Grindelwald glacier, where one of us, in his examination of the bands, was accompanied by Dr. HOOKER, this change in the distribution of the dirt,—the squeezing, lateral drawing act, and bending of the dirt patches below the bottom of the ice-fall,— was especially striking.

Such then appears to be the explanation of the dirt-bands in the cases where we have had an opportunity of observing them. We have not seen those described by Professor FORBES, but the conditions under which he has observed them appear to be similar. An illustration of the explanation just given is furnished by the dirt-bands observed below the "cascade" of the Talèfre. The character of this ice-fall may be inferred from the following words of Professor FORBES, and from the map which accompanies his 'Travels.' "The structure," he says, "assumed by the ice of the Talèfre is extirpated wholly by its precipitous descent to the level of the Glacier de Léchand, where it reappears, or rather is reconstructed out of the broken fragments according to a wholly different scheme." One of the results of this "scheme" would, it is presumed, be a redistribution of the dirt, and the formation of bands in the manner described. Those who consult the map will, however, see dirt-bands marked on the Glacier du Géant also, while no cascade is sketched upon it; but at page 167 of the 'Travels,' Professor FORBES, in referring to this glacier, says, "I am not able to state the exact number of dirt-bands between *the foot of the ice cascade* opposite La Noire and the corner of Trelaporte." Here we are not only informed of the existence of a cascade, but are left to infer that the dirt-bands begin to form at its base, as in the Glacier du Géant, and in those which have come under our own observation. The clean Glacier des Bossons, also, which was referred to by Professor FORBES, in one of his earliest letters, as affording no lodgement to the dirt, possesses its cascade (page 181), and here also we find (page 182) " that the peculiar phenomena of ' *dirt-bands*' on a great scale are not wanting, although from the dazzling whiteness of the ice they may be very easily overlooked." We make these remarks with due reserve, not having yet seen the glaciers referred to.

The explanation just given has been brought to the test of experiment. ABCD, fig. 10, is a wooden trough intended roughly to represent the glacier of the Rhone, the space ACEF being meant for the upper basin. Between EF and GH the trough narrows and represents the precipitous gorge down which the ice tumbles, while the wide space below represents the comparatively level valley below the fall, which is filled with the ice, and constitutes the portion of the glacier seen by travellers descending from the Grimsel or the Furka pass. ACLM is a box with a sluice front, which can be raised so that the fine mud within the box shall flow regularly into the trough, as in the

Fig. 10.

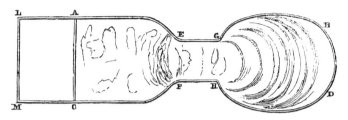

cases already described. The disposition of the trough will be manifest from the section, fig. 11. While the mud was in slow motion downwards, a quantity of dark-coloured

Fig. 11.

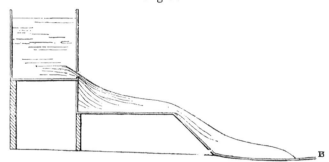

sand was sifted over the space ACEF, so as to represent the débris irregularly scattered over the corresponding surface of the glacier; during the passage of the mud over the brow at EF, and down the subsequent slope, it was hacked irregularly, so as to represent the dislocation of the ice in the glacier. Along the slope this hacking produced an irregular and confused distribution of the sand; but lower down, the patches of dirt and the clean spaces between them gradually assumed grace and symmetry; they were squeezed together longitudinally and drawn out laterally, bending with the convexity downwards in consequence of the speedier flow of the central portions, until finally a system of bands was established which appeared to be an exact miniature of those exhibited by the glacier. On fig. 10 is a sketch of the bands observed upon the surface of the mud, which however falls short of the beauty and symmetry of the original. These experiments have been varied in many ways, with the same general result.

In conclusion we would remark, that our joint observations upon the glaciers of Switzerland extended over a period of a few days only. Guided by the experience of our predecessors, much was seen even in this brief period; but many points of interest first suggested themselves during the subsequent experimental investigation. While, therefore, expressing our trust that the substance of the foregoing memoir will be found in accordance with future observation, we would also express our belief in the necessity of such observation. Indeed the very introduction of the principle of regelation, without which it may be doubted whether the existence of a glacier would be at all possible, opens, in itself, a new field of investigation. This and other questions, introduced in the foregoing pages, must however be discussed with strict reference to the phenomena as Nature presents them. Much might be said even now upon these subjects, but the known liability of the human mind to error when speculation is substituted for observation, renders it safer to wait for more exact knowledge, than to hazard opinions which an imperfect acquaintance with the facts must necessarily render to some extent uncertain.

Royal Institution,
January 1857.